Positive Displacement Pumps

POSITIVE DISPLACEMENT PUMPS

By

Dr. K. M. SRINIVASAN

B.E.(Hons), Ph.D.(USSR)
Dean (R & D), Mechanical Sciences
Kumaraguru College of Technology, Coimbatore
Tamil Nadu

UNIVERSITY SCIENCE PRESS

(An Imprint of Laxmi Publications Pvt. Ltd.)

An ISO 9001:2015 Company

BENGALURU • CHENNAI • GUWAHATI • HYDERABAD
KOCHI • KOLKATA • LUCKNOW • RANCHI
NEW DELHI

Printed and bound in India
First Edition : 2011; *Edition* : 2025
ISBN : 978-93-8085-668-1

Branches

Bengaluru	080-26 75 69 30
Chennai	044-24 34 47 26
Guwahati	0361-254 36 69
Hyderabad	040-27 55 53 83
Kochi	0484-405 13 03
Kolkata	033-40 04 77 79
Lucknow	0522-430 36 13
Ranchi	0651-224 24 64

Published in India by

UNIVERSITY SCIENCE PRESS
(An Imprint of Laxmi Publications Pvt. Ltd.)

An ISO 9001:2015 Company
113, GOLDEN HOUSE, GURUDWARA ROAD, DARYAGANJ,
NEW DELHI - 110002, INDIA

Corporate Office : B-33, Sector-5, Noida,
Gautambuddha Nagar (U.P.) - 201 301, INDIA
Telephone : +91-120-44 46 900
www.laxmipublications.com
info@laxmipublications.com

THIS BOOK is dedicated to
my Mother
Smt. K. T. SAMBOORNAM
and to my Father
Sri. K. MUTHUSAMY PILLAI
as well as
to my professor
Dr. A. A. LOMAKIN
and to my guide
Dr. A. N. PAPIR
Leningrad Polytechnic, Leningrad, K-21 USSR
(at present called as St. Petersburg Polytechnic, St. Petersburg, Russia)
Who brought me to this level.

CONTENTS

PREFACE

It was my long felt ambition to provide a detailed and full information about the theory, design, and analysis of different types of Positive Displacement Pumps. Among the various Positive Displacement Pumps only a few, namely Reciprocating, Radial Piston, Axial Piston, Vane, Gear, Screw, Vortex (Regenerative) and Ejectors and Injectors (Jet pumps) are dealt with in this book, since these pumps occupy major area in the field. I have learned a lot during the period 1959–62 about pumps at PSG College of Technology, Coimbatore – 641 002, while working as Senior Research Assistant for CSIR scheme on Pumps, Turbo charges and flow meters. At the same time, I was also working in Foundry, Pattern Making, Moulding, Production, Testing and Design of different pumps at PSG Industrial Institute, Coimbatore-4 and also during the period 1967 and 1975.

During my Doctorate degree study at Leningrad Polytechnic, Leningrad K-12, USSR (now St. Petersburg Polytechnic, St. Petersburg, Russia,) during 1962 – 1966, Dr. A. A. Lomakin, Dr. A. N. Papir, Dr. Gurioff, Dr . N. Kovaloff, Dr. A. N. Smirnoff, Dr. Staritski, Dr. Gorgidjanyan, Dr. Gutovski were the key professors, who made me to know more about pumps from fundamentals to updated technology. I am very much grateful to Dr. A. A. Lomakin and Dr. A. N. Papir, who were my professor and guide for my doctorate degree course in pumps. Also I have undergone training as a part of my Ph.D. work at Kirov Pump factory and at Leningrad Turbine factory at Leningrad USSR and learned a lot in these factories particularily about the industrial design and the comparison between model and proto type. Also I learned a lot about Dr. Joukoski's work on Propulsion, at Moscow Power Institute, Moscow, USSR. As consultant, for different pump industries in India and abroad, I could understand the field problems. My experience from 1959 till date, has been put up in this book to enable the readers in industries, and in academic area, to design these pumps. Designs examples are also given for a few pumps at the end of each chapter.

I do hope that the reader will be in a position to understand, design, test and analyze pumps, after going through this book. I shall be very much honoured if my book is useful in attaining this. I am grateful to our Chairman Arutselvar Padma Bhushan Dr. N. Mahalingam, Co-Chairman Dr. Krishnaraj Vanavarayer, Vice-chairman Dr. K. Arumugam, Correspondent Sri M. Balasubramaniyam, and the Joint Correspondent Dr. A. Selvakumar for their help in preparing this book. I am also thankful to my wife Smt. S. Nalini, my sons Sri S. Muthuraman, Sri S. Jaganmohan and my daughter Smt. S. Nithyakala, who were very helpful in preparing the manuscript and drawings. I am also thankful to Ms. N. Malarkodi my assistant and Mr. M. Krishnakumar my assistant, who were the key persons to bring the manuscript in the fully typed version with figures and tables.

Constructive suggestions for the improvement of the quality of this book further is always welcome. I earnestly request the readers to send their comments and suggestions either to me direct or through the Publishers.

—Author

CHAPTER 1

RECIPROCATING PUMP

1.1 PRINCIPLE, CONSTRUCTION AND AREA OF APPLICATION

The development of steam turbine, the application of supporting equipments for the main equipments, pumping of viscous liquids such as crude oil, and chemical liquids, which ordinary centrifugal pumps cannot pump, are the main causes for the development of reciprocating pumps.

Reciprocating pumps are classified into two groups as (*a*) piston pumps and (*b*) plunger pumps. Piston pumps can be single acting or double acting. Parallel operation of two or three or more reciprocating pumps or plungers is common. This increases the quantity of flow for the same pressure. Further, in reciprocating pumps and in general in all positive displacement pumps, the pressure is independent of the quantity pumped. Pumping quantity depends only upon speed of the pump and dimensions.

A simple working sketch of a reciprocating pump is illustrated in figure 1.1.

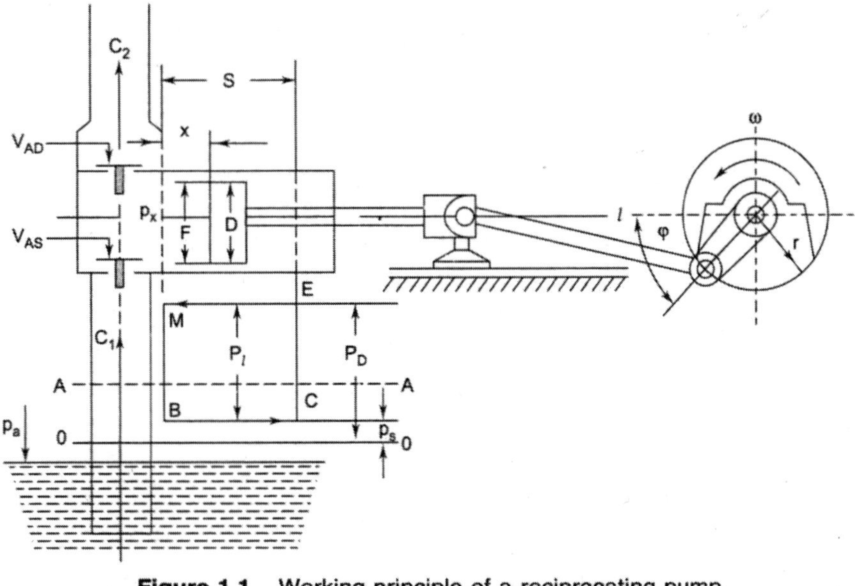

Figure 1.1. Working principle of a reciprocating pump

When the piston moves from the left end of the cylinder towards the right end, due to the rotation of the crank shaft, the pressure p_x goes below atmospheric pressure p_{at}. Due the pressure difference, $\Delta p = p_{at} - p_x$, the suction valve, V_s, opens. Liquid enters from the suction sump to the cylinder through suction pipe, until the piston reaches the inner dead centre, *i.e.*, the right side end of the cylinder, travelling a distance of S, the stroke length of the piston. During this time the crank has rotated half revolution. During the other half of the crank rotation, the piston moves from right side to the left side.

The pressure, p_x, of the liquid in the cylinder increases momentarily, due to the incompressible character of the liquid. Since the delivery pressure and the suction sump pressure are lower than the pressure inside the cylinder, the delivery valve V_{AD} opens and the suction valve V_{AS} closes. The liquid from the cylinder rushes to the delivery tank through the delivery pipe. Thus, one cylinder volume V = (A × S), where A is the area of the piston = $(\pi D^2/4)$, D is the diameter of the piston, and S is the stroke of the piston, of the liquid is transferred from the suction to the delivery during one revolution of the crank.

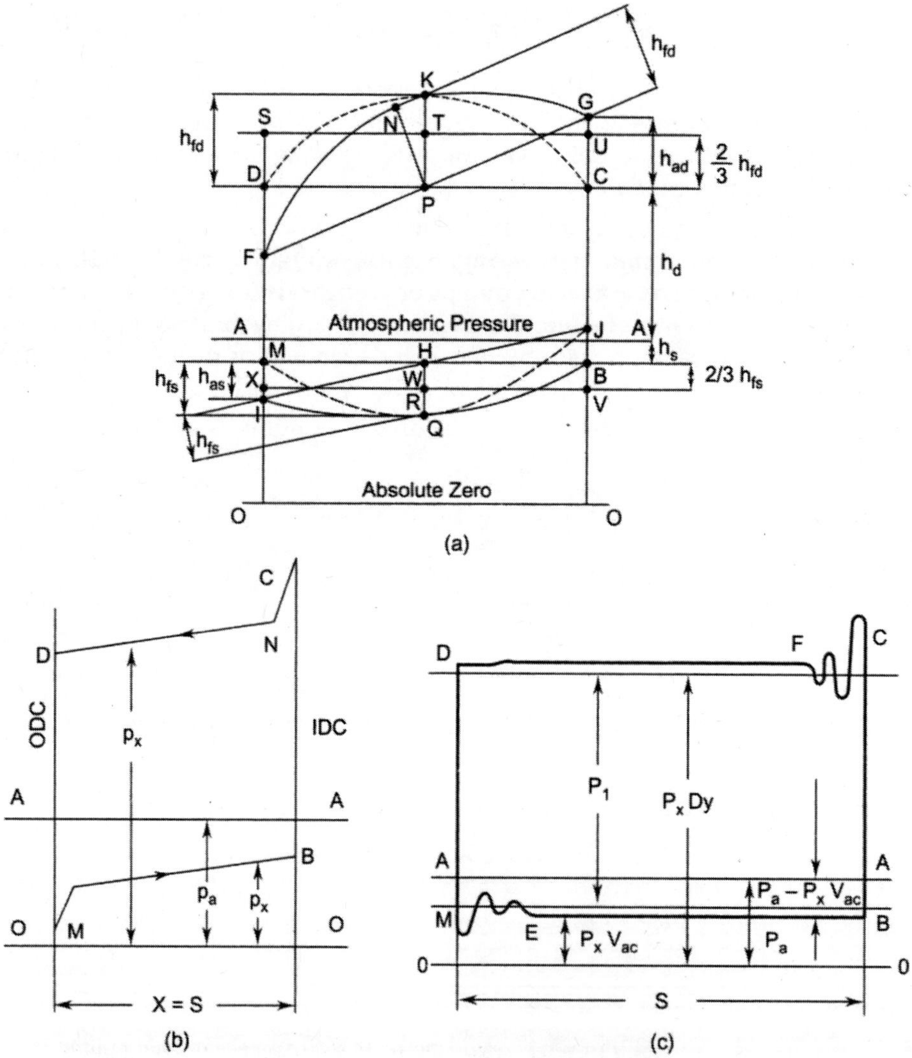

Figure 1.2. Theoretical indicator diagram

The working principle of the reciprocating pump can be easily understood by the indicator diagram MBCDM (Fig. 1.2). The line A – A, in the diagram, indicates the atmospheric pressure p_{at}, and the line O – O, indicates the absolute zero. The suction valve opens at the point M and the liquid enters the cylinder up to the point B. At this point, the piston returns and compresses the liquid. The pressure inside the cylinder momentarily raises from suction pressure h_s at point B to the delivery pressure h_d at point C. Due to sudden increase in the pressure inside the cylinder the piston moves from C to D. At this point the delivery valve opens and the suction valve closes. The liquid starts moving from the cylinder, through the delivery pipe, to the delivery point. This process is indicated by the line CD. At the point D, the pressure inside the cylinder instantaneously drops down to the vacuum (Line D to M). Thus, during one revolution of the crank shaft, the piston moves from M to M through points B, C, D. The line DC = MB is the stroke lengths S of the piston. The pressure p, during suction stroke is p_s and during delivery stroke is p_d.

Volume of the liquid pumped during one stroke or one revolution

$$V = A \times S = (\pi D^2/4) \times S$$

The total volume pumped during n revolutions per minute of the crank is the theoretical quantity

$$Q_{th} = V \times n/60 = A \times S \times n/60 = (\pi D^2/4) \times S \times n/60$$

where V is the cylinder volume = A × S

A is the area of cylinder = $(\pi D^2/4)$, D is the piston diameter, S is the stroke length and n is the speed in rpm.

The pressure rise during this time

$$p = p_d - p_s$$

Expressing the pressure in N/m^2, and the specific weight of the liquid in N/m^3.

The power N_o spent will be

$$N_o = P \times V \times n/60 \text{ watts.}$$

Actual power absorbed will be more due to the friction losses in the real fluid flow, leakage losses through the clearances and mechanical losses in the bearings. The power output N_o will be

$$N_o = \frac{\gamma QH}{\text{const}} = \frac{\gamma \pi D^2 Sn(h_s + h_d)}{4 \times 60 \times \text{const}} \quad ...(1.1)$$

where γ is the specific weight of the pumping fluid at the pumping temperature.

H is the total head $(h_d - (-h_s)) = h_d + h_s$, where h_d is the total delivery head = (p_d/γ), h_s is the total suction head = (p_s/γ), and Q_{th} is the theoretical quantity pumped expressed in m^3/sec, $p = \gamma$ H, the total pumping pressure in N/m^2, W = γ Q, the weight of the pumping liquid in N.

The value of constant will be 1, if the power is expressed in watts and 1000 if the power is expressed in kW. If the efficiency of the pump is η_p and that of the motor is η_m then overall efficiency η of the system will be $\eta = \eta_p \times \eta_m$.

Power input to the pump

$$N_{IP} = \frac{N_o}{\eta_p} \quad ...(1.2(a))$$

Power input to the motor

$$N_I = \frac{N_{IP}}{\eta_m} \quad ...(1.2(b))$$

In many occasions, the delivery and suction pipes are fitted on either side of the cylinder thereby both sides of the piston will be in action. During forward stroke of the piston, front side of the piston acts as delivery stroke and pushes water to the delivery pipe, while the rear end of the piston, acts as suction stroke, and water is sucked from the suction sump. When the piston moves backward, the suction and delivery strokes are inter changed. Due to this, the quantity of flow given by the pump during each revolution will be nearly doubled *i.e.*,

$$Q_{th} = \frac{2\pi D^2 Sn}{4 \times 60} \quad ...(1.3)$$

These types of pumps are called Double Acting Reciprocating pumps, while the former types are called Single Acting Reciprocating pumps.

1.2 PULSATING FLOW IN THE PUMP AND ITS EFFECTS

When the crank moves from M to B, the piston moves from ODC to IDC, the stroke of the piston will be equal to twice the crank radius *i.e.*, $S = 2r$ where r is the crank radius. Assuming that, the liquid in the cylinder also moves at the same velocity and acceleration as that of the piston, at any time 't' the distance moved by the piston 'x' will be equal to

$$x = r - r\cos\phi$$

where ϕ is the angle of rotation of the crank from the dead centre.

Since $\phi = \omega t$, where ω is the angular velocity of the crank

$$= (2\pi n/60), \frac{2\pi n}{60} \times t.$$

Velocity of the piston, and hence the velocity 'V' of the liquid inside the cylinder, is equal to

$$V = [dx/dt] = d(r - r\cos\omega t)/dt = \omega r \sin\omega t = \omega r \sin\phi.$$

Consider the liquid in the pipe (suction or delivery). Since the quantity of flow passing through the pipe is same as the quantity coming out of the cylinder,

$$Q = a \times v = A \times V,$$

where a and A are area of the pipe and cylinder respectively and v and V are the velocities of the liquid in pipe and cylinder respectively. The velocity of liquid in the pipe 'v' will then be equal to

$$v = \frac{VA}{a} = \frac{A}{a}\omega r \sin\phi = \frac{A}{a}\omega r \sin\omega t \quad ...(1.4)$$

The quantity of flow passing through the pipe is same as 'Q_{th}' and will be equal to = $a.v$ = A ωr sin ωt. This equation indicates that the quantity is not constant as in centrifugal pump or in any rotary pump; instead it is pulsating in simple harmonic motion. This can as well be represented in the form of a diagram (Fig.1.2). At any time, when the crank moves through an angle $\phi = f(\omega t)$, the quantity (Q_{th}) entering the pipe will be as indicated in the diagram (equation1.4). When $\phi = 0°$ or 180°, no quantity flows through the pipe, whereas when $\phi = 90°$, the maximum quantity flows through the pipe. This indicates that the flow is not uniform in the pipe, instead it is pulsating flow. This fluctuation must be avoided. This is done by fitting an airvessel very close to the pump, at suction side immediately before the suction valve. The liquid enters the air vessel and then reaches the pump. Correspondingly at the delivery side, the liquid enters the air vessel which is fitted immediately after the delivery valve and then goes to the pipe.

The air vessel stores the excess quantity coming out, of the pump and delivers a constant quantity to the pipe at all times, thereby uniform flow is maintained in the pipe at all times. Similar action takes place in suction pipe also. Liquid enters the suction air vessel from the suction pipe and than enters the cylinder through suction valve. Uniform flow prevails in pipe from suction sump to air vessel. Fluctuated flow prevails from the cylinder to the air vessel.

1.3 ACCELERATION HEAD AND ITS EFFECT

1.3.1 Pressure Variations and Pressure Head

The reciprocating motion of the piston or plunger is a simple harmonic motion, which produces acceleration at the beginning of the stroke and retardation at the end of the stroke. Correspondingly this acceleration and retardation are transmitted to the pumping fluid, which results in pressure variations. Assuming simple harmonic motion, the acceleration of the piston at any time 't' will be

$$a_p = dV/dt = d(\omega r \sin\phi)/dt = \omega^2 r \cos\phi$$
$$= \omega^2 r \cos \omega t$$

Corresponding acceleration in pipe, 'a' will be

$$= d[(A/a)\,\omega r \sin\phi]/dt = (A/a)\,\omega^2 r \cos\phi/dt = (A/a)\,\omega^2 r \cos\omega t.$$

If the length of the pipe (delivery or suction) is l the mass of water m, flowing at any time will be

$$m = (\gamma a l/g),$$

where g is the acceleration due to gravity and γ is the specific weight of the liquid.

Force (F) developed due to fluid acceleration, (as per Newton's law) is equal to

$$F = ma = \frac{\gamma a l}{g}\frac{A}{a}\,\omega^2 r \cos\phi \qquad ...(1.5)$$

Pressure (p_a) developed due to the fluid acceleration will be equal to

$$p_a = \frac{F}{a} = \frac{\gamma l}{g}\frac{A}{a}\ \omega^2 r \cos\phi \qquad ...(1.6)$$

Pressure head (h_a) developed due to the fluid acceleration will be equal to

$$h_a = \frac{p_a}{\gamma} = \frac{l}{g}\frac{A}{a}\ \omega^2 r \cos\phi \qquad ...(1.7(a))$$

For suction stroke $$h_{as} = \frac{l_s}{g}\frac{A}{a_s}\ \omega^2 r \cos\phi \qquad ...(1.7(b))$$

For delivery stroke $$h_{ad} = \frac{l_d}{g}\frac{A}{a_d}\ \omega^2 r \cos\phi \qquad ...(1.7(c))$$

Calculating the values of h_{as} and h_{ad} for different positions of crankshaft, for the suction, when, $\phi = 0°$, at the beginning of the suction stroke as well as $\phi = 180°$, at the end of the suction stroke.

$$h_{as} = \pm \frac{l_s}{g}\frac{A}{a_s}\ \omega^2 r, \qquad ...(1.8)$$

since, $\cos 0° = 1$; and $\cos 180° = -1$

When $\phi = 90°$ (at the middle of the suction stroke) $h_{as} = 0$, since, $\cos 90° = 0$

For the delivery, in the similar way, when $\phi = 0°$ and $\phi = 180°$ (at the beginning and at the end of the delivery stroke)

$$h_{ad} = \pm \frac{l_d}{g}\frac{A}{a_d}\ \omega^2 r \qquad ...(1.9)$$

When $\phi = 90°$ (at the middle of the delivery stroke), $h_{ad} = 0$

Different values of h_{as} and h_{ad} are marked in the diagram of work done (Fig. 1.2). Area of the diagram MBCD changes to area FIJG. But area DFE = area ECG and area MIH = area HBJ. Hence area MBCD = area FIJG which indicates that work done by the pump does not change due to simple harmonic motion of the piston, correspondingly, due to acceleration head.

1.3.2 Importance of Acceleration Head

The total head at the beginning of the suction stroke, $H_s = h_s + h_{as}$. Also total head at the end of delivery stroke, $H_d = h_d - h_{ad}$. If both these values exceed 8 m of vacuum, the liquid will no longer remain in liquid form, instead it will be in vapour form, which induces cavitation, separation of flow, vibration and noise. Hence while selecting the pipe size and length for suction and delivery, care should be taken to minimize the acceleration head which in turn reduces the total head at the beginning of the suction stroke or at the end of the delivery stroke. Larger diameter and smaller lengths of the pipe reduce the acceleration head.

1.4 EFFECT OF FRICTION AND ITS CALCULATION

Friction head 'h_f' is always calculated by Darcy – Weisbach equation and can be written as,

$$h_f = \frac{flv^2}{2gd},$$

where f is the friction coefficient. Substituting the value of velocity in pipe v, from the equation 1.4,

$$h_f = \frac{fl}{2gd}\left[\frac{A}{a}\omega r \sin\phi\right]^2 \quad ...(1.10(a))$$

For the suction,

$$h_{fs} = \frac{fl_s}{2gd_s}\left[\frac{A}{a_s}\omega r \sin\phi\right]^2 \quad ...(1.10(b))$$

For the delivery,

$$h_{fd} = \frac{fl_d}{2gd_d}\left[\frac{A}{a_d}\omega r \sin\phi\right]^2 \quad ...(1.10(c))$$

Referring to equation 1.10, it is evident that the variation between ϕ and h_f is parabolic, ($h_f \propto \sin^2\phi$). The maximum value of friction head is at the middle *i.e.*, when $\phi = 90°$. Since $\sin 90° = 1$, and h_{fs} is equal to

For the suction stroke

$$h_{fs} = \frac{fl_s}{2gd_s}\left[\frac{A}{a_s}\omega r\right]^2 \quad ...(1.11)$$

For the delivery stroke

$$h_{fd} = \frac{fl_d}{2gd_d}\left[\frac{A}{a_d}\omega r\right]^2 \quad ...(1.12)$$

This is shown in Fig. 1.2. From the diagram, it is evident that area of the diagram is now increased. Area "FIJG" is changed to area FILJGKF. This increase is due to the added area FKGEF at the delivery and area ILJH at the suction. Maximum ordinate h_{fd} for delivery and h_{fs} for suction, are occurring at the center of delivery and suction strokes respectively.

In order to avoid cavitation, separation of flow, noise and vibration, it is essential to design the total suction head $H_s = (h_s + h_{fs})$ should not be more than 8 m of vacuum. This can be achieved by selecting proper diameter and length of the suction pipe.

Area of parabola

$$\text{FKGEF} = \frac{2}{3}\,\text{KE} \times \text{FG} = \frac{2}{3}\,\text{KE} \times \text{BC} = \frac{2}{3}\,h_{fd} \times \text{BC}$$

and Area

$$\text{ILJH} = \frac{2}{3}\,\text{HL} \times \text{IJ} = \frac{2}{3}\,\text{HL} \times \text{MD} = \frac{2}{3}\,h_{fs} \times \text{MD}$$

Based on this, the diagram can be redrawn. Since area FGJI = area MBCD, area FKG = area DKC and area IHJL = area MLB. Total work done by the pump due to static height, acceleration head, and friction head will be equal to the area of the diagram DKCBLMD. The output from the pump

$$N_o = \frac{\gamma Q_a H_a}{\text{const}} = \frac{\gamma}{\text{const}} \times \frac{\pi D^2}{4} S \frac{n}{60} \times \left[h_s + \frac{2}{3} h_{fs} + h_d + \frac{2}{3} h_{fd} \right] \quad ...(1.13)$$

The power input $N_I = \dfrac{N_o}{\eta}$ ...(1.14)

1.5 PRESSURE UNDER THE PISTON INSIDE THE CYLINDER, DURING SUCTION AND DELIVERY STROKES

Suction characteristics of the pump depend upon the pressure at the bottom side of the piston inside the cylinder during suction, which is depending upon the suction head, speed and construction of the pump. Unsteadiness in flow is determined by the number of pumps running in parallel. In multistage pumps, the unsteadiness in the pipe is reduced to a considerable extent. But the flow within the cylinder is always non-uniform.

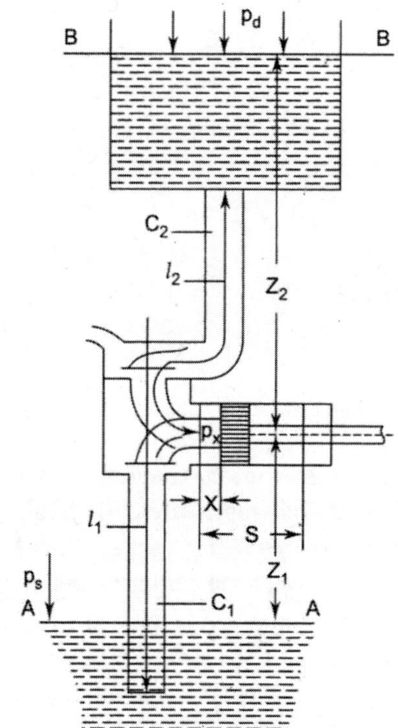

Figure 1.3. Working principle of the reciprocating pump

1.5.1 Pressure Inside the Cylinder During Suction Stroke

During suction stroke the piston moves from outer dead centre to inner dead centre. Applying Bernoulli's equation between A-A and the piston centre, *i.e.*, shaft axis. (Fig. 1.3)

$$\frac{P_{at}}{\gamma} = \frac{P_x}{\gamma} + Z_1 + \frac{v_1^2}{2g} + \Sigma h_{fs} + \frac{1}{g} \int_0^{l_1 + x} \frac{\partial v}{\partial t} dl \quad ...(1.15)$$

The last term in the above equation indicates the inertia head, which is the energy required to move the mass inside the pipe and the cylinder, for a length of '$l + x$' with acceleration of $\frac{\partial v}{\partial t} = \frac{dv}{dt}$, since v is the function of t only.

$$\frac{dv}{dt} = \omega^2 r \cos\phi \qquad ...(1.16)$$

Assuming the pipe and the piston have same diameter and combining the above two equations and rearranging

$$\frac{p_x}{\gamma} = \frac{p_a}{\gamma} - \left(Z_1 + \frac{v^2}{2g} + \Sigma h_{fs} + \frac{l_1 + x}{g} \omega^2 r \cos\phi \right) \qquad ...(1.17)$$

The flow velocity inside the cylinder is very small, hence neglected. Since the losses in the valves, due to sudden closing and opening and the losses in the pipe are considerable, h_{fs} (or) h_{fd} is always taken into account.

The inertia force and corresponding inertia head depends upon the piston movement. when $\phi = 0$, $\cos\phi = 1$, it will be $\frac{l_1 + x}{g}\omega^2 r$ and when $\phi = 90°$, $\cos\phi = 0$, this will be zero.

Correspondingly the inertia head is maximum at the beginning of the suction stroke ($x = 0$). The pressure is maximum at the bottom of the piston. This value gradually decreases, when ϕ changes from 0 to 90° and then increases in the negative direction, when ϕ changes from 90° to 180°. Figure 1.4 illustrates the change of $p_x = f(x)$, and is represented by the line MPB. In figure 1.2(c), the line 0 – 0 is the absolute vacuum line; line A – A is the atmospheric pressure line. The line MPB indicates the change as per the equation (1.15). The line MP represents the loss due to the valve opening at the beginning of the suction stroke. Hence cavitation and flow separation can take place only at the beginning of the suction stroke, as far as inertia force is concerned.

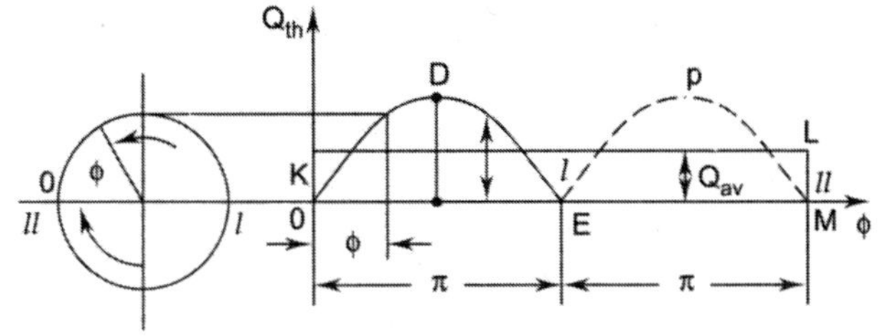

(a) One pump in operation

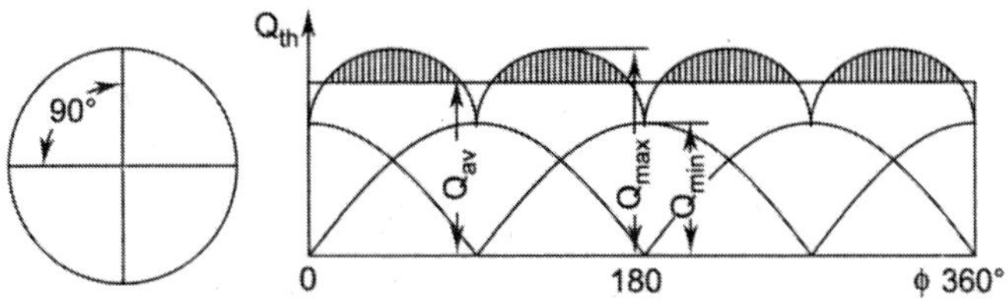

(b) Two pump in operation

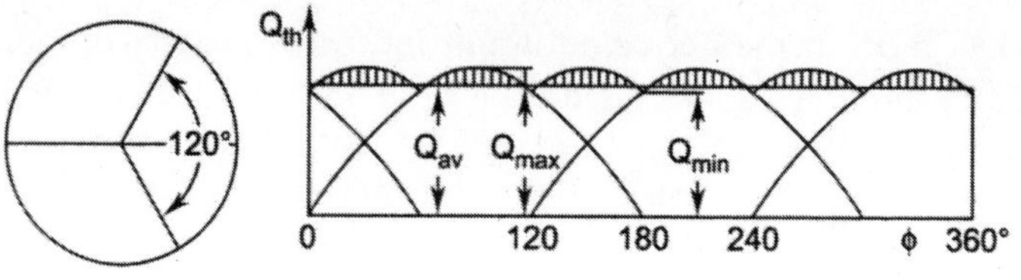

(c) Three pumps in operation

Variation of Q_{th} with respect to angle of rotation ϕ

Figure 1.4. Indicator diagram with the effect of valve opening

Due to non-uniform motion of the piston, fluid also moves in the same non-uniform motion. The equation for ideal fluid flow can be written as

$$\frac{\partial}{\partial l}\left(\frac{p}{\gamma}+x+\frac{v_s^2}{2g}\right)+\frac{1}{g}\frac{\partial v}{\partial t}=0 \quad \text{...(1.18)}$$

The first part of the above equation gives the change in the fluid energy along the flow path and second part gives the inertia force. In real fluid, the flow is affected by the reduction of the total energy due to fluid friction, The above equation is written

$$\frac{\partial}{\partial l}\left(\frac{p}{\gamma}+x+\frac{v^2}{2g}\right)+\frac{1}{g}\frac{\partial v}{\partial t}+h_{fs}=0 \quad \text{...(1.19)}$$

where, h_{fs} is the loss due to viscosity of the liquid and inertia of the valve. This loss cannot be considered as a function of length l. It is always expressed as total loss. Multiplying the above equation by length dl and integrating the above equation with respect to l, between (suction sump) $l = 0$ and(the pump) $l = l$

$$\int_{l=0}^{l=lx}\frac{\partial}{\partial l}\left(\frac{p}{r}+x+\frac{v^2}{2g}\right)dl+\frac{1}{g}\int_{l=0}^{l=lx}\frac{\partial v}{\partial t}dl+\sum_{l=0}^{l=lx}h_{fs}\,dl=0 \quad \text{...(1.20)}$$

The velocity V_o at suction sump is zero and inside the cylinder, the fluid moves with the same velocity as that of the cylinder 'u', the first integral can be written as

$$\int_0^{lx}\frac{\partial}{\partial l}\left(x+\frac{p}{\gamma}+\frac{v^2}{2g}\right)dl = Z_1+x+\frac{p_{x1}}{\gamma}-\frac{p_0}{\gamma}+\frac{u_x^2}{2g} \quad \text{...(1.21)}$$

where p_o is pressure at suction sump.

From the continuity equation $\qquad v_s\frac{\pi d_s^2}{4} = u\frac{\pi D^2}{4}$

where v_s and d_s are the fluid velocity and diameter in suction pipe and u and D are the fluid velocity inside the cylinder and diameter of the cylinder,

$$v_s=\frac{uA}{a_s}=\frac{uD^2}{d_s^2} \text{ and } \frac{\partial v}{\partial t}=\frac{\partial u}{\partial t}\frac{A}{a_s}$$

Considering each part of suction line from sump to cylinder, and since $\frac{\partial u}{\partial t}$ is independent of length 'l, the second part of the equation (1.20) can be written as

$$\frac{1}{g}\int_0^{lx}\frac{\partial v}{\partial t}\,dl = \frac{1}{g}\cdot\frac{\partial u}{\partial t}\int_0^{lx}\frac{A}{a}\,dl = \frac{1}{g}\frac{\partial u}{\partial t}\left[\frac{A}{a_1}l_1+\frac{A}{a_2}l_2+\ldots+\frac{A}{a_x}l_x = \frac{1}{g}\frac{\partial u}{\partial t}\left[l_s+x\right]\right] \quad \text{...(1.22)}$$

The third part of the equation (1.20) indicates the sum of hydraulic losses due to friction, major and minor losses in pipe and losses in valves and cylinder. Since $v_{si} = \frac{A}{a_{si}}.u_x$

$$\sum_0^n f\frac{v_{si}^2}{2g}\cdot\frac{l_{si}}{d_{si}} = \frac{u_x^2}{2g}\sum_0^n f\frac{l_{si}}{d_{si}}\left[\frac{A}{a_{si}}\right]^2 \quad \text{...(1.23)}$$

Minor losses can be

$$\sum_{k=0}^{k=m} k\frac{v_k^2}{2g} = \sum_0^m k\cdot\frac{u_x^2}{2g}\cdot\left[\frac{A}{a_k}\right]^2 = \frac{u_x^2}{2g}\sum_0^m k\cdot\left[\frac{A}{a_k}\right]^2 \quad \text{...(1.24)}$$

Total losses h_{fos}, can be expressed as

$$\sum_0^{i_k} h_{fsi}\,dl = \sum_0^n f_i\frac{l_i}{d_i}\left[\frac{A}{a_z}\right]^2+\sum_0^m K\left[\frac{A}{a_x}\right]^2\frac{u_x^2}{2g} \quad \text{...(1.25)}$$

f_1 denotes the coefficient of friction for the suction pipe of length I_s and K, the coefficient for minor losses. Taking $f_1 + K = f_s$, combining all the three equations (1.21, 1.22, 1.25).

$$z_1+x+\frac{p_{ix}}{\gamma}-\frac{p_0}{\gamma}+\frac{u_x^2}{2g}+\frac{1}{g}\frac{\partial u}{\partial t}(l_s+x)+f_s\frac{u_x^2}{2g}+h_{fos}=0 \quad \text{...(1.26)}$$

from which $$\frac{p_{ix}}{\gamma}=\frac{p_0}{\gamma}-\left[z_1+x+\frac{1}{g}\frac{\partial u}{\partial t}(l_s+x)+(1+f_s)\frac{u_x^2}{2g}+h_{fos}\right] \quad \text{...(1.27)}$$

From (Fig. 1.1), $x = r(1-\cos\phi)$.

And $\left(\cos\phi = 1-\frac{x}{r}\right)$ since $u = \omega r\sin\phi$

$$\frac{u^2}{2g} = \frac{\omega^2 r^2}{2g}\sin^2\phi = \frac{\omega^2 r^2}{2g}\left(1-\cos^2\phi\right)$$

$$= \frac{\omega^2 r^2}{2g}\left[1-\left(1-\frac{x}{r}\right)^2\right] = \frac{\omega^2 r^2}{2g}\left[\frac{2x}{r}-\left(\frac{x}{r}\right)^2\right] \quad \text{...(1.28)}$$

$$\therefore \quad \frac{\partial u}{\partial t} = \omega^2 r\cos\phi = \omega^2 r\left[1-\left(\frac{x}{r}\right)\right]$$

So, $$\frac{p_x}{\gamma}=\frac{p_0}{\gamma}-\left[Z_s+x+\frac{l_s}{g}\omega^2(r-x)+\frac{x}{g}\omega^2(r-x)+(1+f_s)\frac{\omega^2 r^2}{2g}\left(\frac{2x}{r}-\left(\frac{x}{r}\right)^2\right)+h_{fos}\right] \quad \text{...(1.29)}$$

Equation (1.29) can be used to study the pressure at the piston at any instant of its position. Fig. 1.5(*b*) gives the variation of each element in the equation(1.29), as a function of *x i.e.*, $\frac{p_{xs}}{\gamma} = f(x)$.

Curve No. 1 is p_o/γ, the sump pressure which is always constant.

Curve No. 2 is $(Z_s + x)$, where Z_s is constant, but x changes and hence inclined.

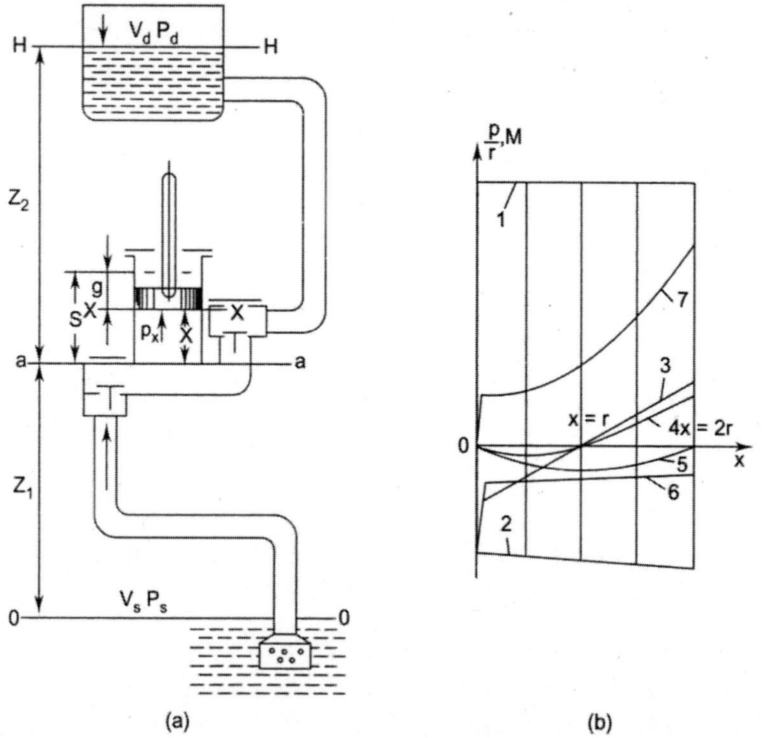

Figure 1.5. Pressure variations inside the cylinder during suction stroke

Curve No. 3 is $\frac{l_s}{g}\omega^2(r-x)$. This value will be $-\frac{l_s}{g}\omega^2 r$ when $x = 0$ ($\phi = 0°$) and 0 when $x = r$, ($\phi = 90°$) and $= +\frac{l_s}{g}\omega^2 r$ when $x = 2r$ ($\phi = 180°$). It changes from negative at the beginning of the stroke, zero at the middle of the stroke and positive at the end of the suction stroke.

Curve No. 4 is $\frac{x}{g}\omega^2(r-x)$ which represents the inertia force of the liquid column in the cylinder. This value is equal to zero when $x = 0$, *i.e.*, at the beginning of the stroke, equal to $-\frac{\omega^2 r^2}{4g}$ when $x = r/2$, again is equal to zero when $x = r$ and is equal to $+\frac{\omega^2 r^2}{4g}$, when $x = 2r$.

Curve No. 5 is $(1 + f_s)\frac{\omega^2 r^2}{2g}\left[\frac{2x}{r} - \left[\frac{x}{r}\right]^2\right]$, and represents the hydraulic friction losses in pipe and in cylinder. This is maximum when $x = r$.

Curve No. 6 is the losses in the valve h_{trs} due to inertia force.

Curve No. 7 indicates sum of all losses, which represent actual pressure inside the cylinder during suction stroke.

From the graph, it is evident that minimum pressure *i.e.*, maximum vacuum exists at the beginning of the suction stroke, when $x = 0$. However, if the suction losses are higher due to the site conditions, such as higher length, lower diameter, too many fittings, higher speed etc., minimum pressure or maximum vacuum may occur at the middle of the stroke, *i.e.*, when $x = r$.

Frictional coefficient does not influence, at the beginning of the stroke, since fluid velocity is zero. For normal working of the pump, the pressure inside the cylinder must be above vapour pressure, p_{vp}, *i.e.*, $p_{smin} > p_{vp}$, otherwise vapour bubbles will form, inside the cylinder and cavitation starts.

Applying the condition $x = 0$ and $p_s = p_{vp}$ the equation(1.19)

$$z_s = \frac{p_0}{\gamma} - \left[\frac{p_{vp}}{\gamma} + \frac{l_s}{g}\omega^2 r + h_{fvs}\right] \qquad \text{...(1.30)}$$

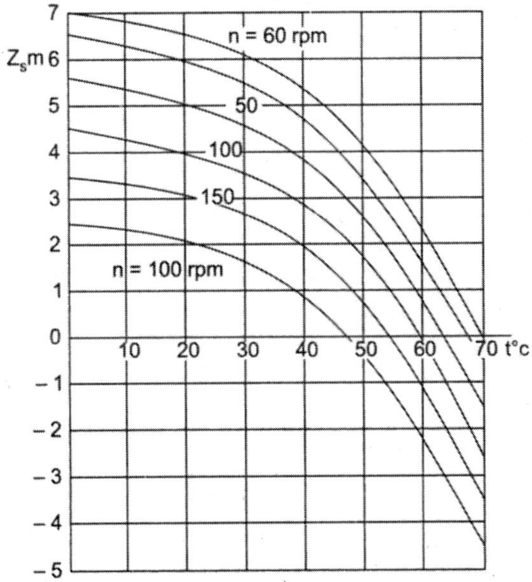

Figure 1.6. Variation of suction lift $Z_s = f(t°C, n)$

Figure 1.6 illustrates, the allowable suction lift Z_s with respect to the temperature, and the allowable speed of the pump. Solving equation (1.30) for ω, in order that $p_s > p_{vp}$

$$\omega = \sqrt{\frac{g\left(\frac{p_0}{\gamma} - \frac{p_{vp}}{\gamma} - z_s - h_{fvs}\right)}{rl_s}} \qquad \text{...(1.31)}$$

If the connecting piston rod is taken into account $r/L = 1/5$, where L is the length of the piston rod.

$$\omega = \sqrt{\frac{g\left(\frac{p_0}{\gamma} - \frac{p_{vp}}{\gamma} - z_s - h_{fvs}\right)}{1.2\, l_s.r}}$$

The speed

$$n \le \sqrt{\frac{150\, g\left(\frac{p_0}{\gamma} - \frac{p_{vp}}{\gamma} - z_s - h_{fvs}\right)}{S.l_s.}} \qquad ...(1.32)$$

Reduction in static height, increase in diameter, reduction in pipe length, reduction in fittings will definitely improve suction conditions and avoid cavitation. Under special conditions, the suction sump is closed and suction tank pressure is increased. Reduction in speed is also recommended in some cases.

1.5.2 Pressure in the Cylinder During Delivery Stroke

The pressure at delivery also changes in the same manner as that of suction. Writing down the Bernoulli's equation for the delivery (figure 1.5(*a*).

$$\frac{p_x}{\gamma} + \frac{v_2^2}{2g} = \frac{p_d}{\gamma} + Z_2 + \sum h_{fd} + \frac{1}{g}\int_0^{l_2+x} \frac{\partial v}{\partial t}.dl \qquad ...(1.33)$$

or

$$\frac{p_x}{\gamma} = \frac{p_d}{\gamma} - \frac{v_2^2}{2g} + Z_2 + \sum h_{fd} + \frac{l_2+x}{g}\omega^2 r\cos\phi \qquad ...(1.34)$$

In figure 1.2(*c*) line CND indicates the delivery process and line NC indicates the effect of opening of the delivery valve. The theoretical indicator diagram is given by the line MPBCNDM in figure 1.2(*c*) which is different from the ideal indicator diagram MBCD (figure 1.2(*a*). From the equation it is found that the pressure at the bottom of the piston, inside the cylinder is changing. When the delivery pressure p_d is smaller and pipe length l_2 is larger, the pressure at the end of the delivery stroke (point K) can go below atmospheric pressure *i.e.*, can be under vacuum. As similar in suction stroke, in delivery stroke also.

When $\phi = 0°$, *i.e.*, at the beginning of the delivery stroke

$$\frac{p_x}{\gamma} = \frac{p_d}{\gamma} + z_2 + \sum h_{fd} + \frac{l_2+x}{g}\omega^2 r \qquad ...(1.35)$$

When $\phi = 180°$, *i.e.*, at the end of the delivery stroke

$$\frac{p_x}{\gamma} = \frac{p_d}{\gamma} + z_2 + \sum h_{fd} - \frac{l_2+x}{g}\omega^2 r \qquad ...(1.36)$$

Where $\sum h_{fd}$ is the loss in the delivery pipe and in the delivery valve. Piston is accelerated during the period 0 to 90° and is decelerated during 90° to 180°. Correspondingly the liquid also gets accelerated during I[st] half of the stroke, but does not decelerate as that of piston in the IInd half, with the result the liquid separates from the piston, which reduces the pressure p_x'. The liquid moves purely due to its own inertia, possessing. Due to the high delivery head and higher friction losses the liquid no longer moves, instead the liquid flows from the delivery pipe into the piston in the reverse direction. This results in hammer blow on the piston. If the pressure, during this time, is equal to suction pressure, the suction valve opens and the liquid starts entering the suction pipe also. This is the error flow and affects the normal flow. All these, yield to the reduction of suction height, inception of cavitation, water hammer, varying pressure etc., which can happen in single and double acting pumps. If there are four and more pumps running together, this action is less and almost a uniform flow exists ($\delta \approx 1.05 - 1.11$), where δ is the fluctuation of coefficient.

Assuming that the piston is located at a distance x from the initial dead centre of the delivery stroke and taking 'y' as the remaining distance to the fluid dead centre at the end of the delivery stroke = $x + y = S$, $y = S - x$. $dy = -dx$

As followed for the suction stroke, writing down the equation for the motion of the piston for a length dl and integrating it from $l = l_x$ to $l = l_d$

$$\int_{lx}^{ld} \frac{\partial}{\partial l}\left(\frac{p}{r} + z + \frac{v^2}{2g}\right) dl + \frac{1}{g}\int_{lx}^{ld} \frac{\partial v}{\partial t}\, dl + \sum_{l=lx}^{l=ld} h_{fdi} dl = 0 \qquad \text{...(1.37)}$$

As done for the suction stroke, a similar equation can be written for delivery as

$$\frac{p_{xd}}{\gamma} = \frac{p_d}{\gamma} + Z_d - x + \frac{v_d^2}{2g} + (f_d - 1)\frac{u_x^2}{2g} + \frac{1}{g}(l_d + x)\frac{\partial u}{\partial t} + h_{fod} \qquad \text{...(1.38)}$$

where p_d and v_d are the pressure and velocity in delivery tank.

Since $x = S - y$, $y = r - r\cos\phi$, $\cos = 1 - \dfrac{y}{r}$

$$u_x = \frac{\partial y}{\partial t} = \omega r \sin\phi,\ \frac{u_x^2}{2g} = \frac{\omega^2 r^2}{2g}\left(\frac{2y}{r} - \left(\frac{y}{r}\right)^2\right),\ \frac{\partial u}{\partial t} = \omega^2 r\left(1 - \frac{y}{r}\right) \qquad \text{...(1.39)}$$

A graph can be drawn between $\dfrac{p_x}{\gamma} = f(y)$ (figure 1.7), where in

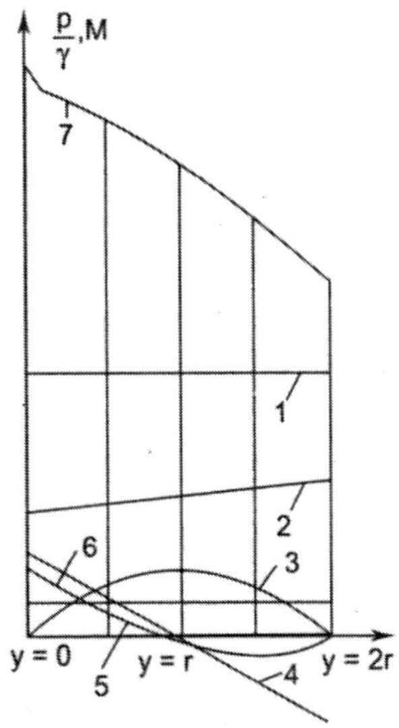

Figure 1.7. Variation of pressure inside the cylinder during delivery stroke

Curve No. 1 represents delivery head $\frac{p_d}{\gamma}$, which is constant.

Curve No. 2 represents potential energy (or) the delivery height $Z_d - x = Z_d - S + y$.
Curve No. 3 represents hydraulic loss in delivery line

$$(f_d - 1)\frac{u^2}{2g} = (f_d - 1)\frac{\omega^2 r^2}{2g}\left[\frac{2y}{r} - \frac{y^2}{r^2}\right]. \quad \text{...(1.40)}$$

Curve No. 4 represents inertia force of liquid column in delivery line

$$\frac{1}{g} l_d \frac{\partial u}{\partial t} = \frac{1}{g} l_d\, \omega^2 r\left(1 - \frac{y}{r}\right). \quad \text{...(1.41)}$$

Curve No. 5 represents inertia force of the liquid in cylinder

$$\frac{1}{g} x \frac{\partial u}{\partial t} = \frac{1}{g}(S - y)\omega^2 r\left(1 - \frac{y}{r}\right). \quad \text{...(1.42)}$$

Curve No. 6 represents the delivery valve losses h_{fvd}.

Curve No. 7 is the total sum of all which is $\frac{p_{xd}}{\gamma} = f(y)$.

Minimum pressure exists when $y = S$, *i.e.*, at the end of delivery stroke. If Z_d, p_d are smaller p_{x2} will be minimum. So this pressure must be greater than the vapour pressure $p_{x2} > p_{vp}$.

Under low pressure p_{x2} prevailing inside the cylinder, flow separation from the piston, easily opening of suction valve, shock losses due to sudden impact of piston on the fluid will happen

Taking $$\frac{p_d}{\gamma} \geq \frac{p_{vp}}{\gamma}, \omega = \frac{2\pi n}{60}, y = S$$

from equation (1.38), the allowable speed n_{max} will be

$$n_{max} \leq \sqrt{\frac{150g\left(\frac{p_d - p_{vp}}{\gamma} + z_d + h_{fvd}\right)}{S.l_d}} \quad \text{...(1.43)}$$

The most common feature is the flow separation from the piston. Reduction of length l_d, increase in diameter of the delivery pipe d_d, speed reduction ω, reduction of crank radius r increasing the delivery tank pressure p_d, provision of air vessel close to the pump will reduce cavitation due to reduced pressure inside the cylinder.

1.6 AIR VESSEL

1.6.1 Estimation of Quantity of Flow Into or Out from the Air Vessel

Excess quantity entering into or leaving the air vessel at any time is

$$Q_C - Q_P = A\,\omega r \sin \omega t - \frac{\pi d^2}{4} v \qquad ...(1.44)$$

where Q_C is the quantity from the cylinder of the pump and Q_P is the quantity flowing through the pipe (If $Q_C > Q_P$ *i.e.*, +ve, excess liquid enters the air vessel and if $Q_C < Q_P$ *i.e.*, –ve, liquid leaves air vessel, in order to maintain a uniform flow at the pipe. The air vessel acts as a source and sinks for the supply of water between the cylinder and the pipe. The air vessel has to be designed to meet the capacity which is maximum when $\phi = 90°$.

1.6.2 Suction

During suction stroke, when the piston moves from ODC to IDC, the liquid raises through a height of (Z_s) Z_1 in suction pipe with a velocity v_1 and gradually fills up the suction air vessel. The pressure inside the air vessel p_{1k}, is less than the suction sump pressure p_s (= p_{at}, point M in Fig 1.2), but higher than the pressure inside the cylinder p_{x1}. So, simultaneously the liquid, from air vessel goes to the cylinder through the suction pipe. The liquid level, in the air vessel *k-k* gradually goes down and pressure, p_{1k} also reduces, which makes the liquid from suction sump to enter air vessel through suction pipe, l'' . The air vessel pressure p_{1k} and the level *k-k* (or) the height Z″ goes on changing. The flow through the pipe length l'', also fluctuates as similar to the non-uniform flow prevailing in cylinder for the distance '*x*'. Writing the equation

$$\frac{p_x}{\gamma} = \frac{p_{1k}}{\gamma} - \left(Z_1'' + \frac{v_1^2}{2g} + \sum h_{fs}'' + \frac{l_1'' + x}{g}\,\omega^2 r \cos\phi \right) \qquad ...(1.45)$$

Where $\sum h_{fs}''$ is the loss due to friction in pipe length l'', mostly due to valve opening and closing. Writing down the Bernoulli's equation between A–A and k–k

$$\frac{p_a}{\gamma} = \frac{p_{1k}}{\gamma} + Z_1' + \sum h_{fs}' \qquad ...(1.46)$$

Combining the above two equations

$$\frac{p_x}{\gamma} = \frac{p_a}{\gamma} - \left(Z_1 + \frac{v_{ap}^2}{2g} + \sum h_f + \frac{l_1'' + x}{g}\,\omega^2 r \cos\phi \right) \qquad ...(1.47)$$

where $\quad Z_1 = Z_1' + Z_1''$ and $\sum h_{fs} = \sum h_{fs}' + \sum h_{fs}''$

Comparing these two equations, (1.17) and (1.47), the inertia head is reduced considerably due to air vessel. The unsteady and non-uniform flow in suction pipe can be reduced to a maximum when the air vessel is fitted very close to the pump and the length l_1'' is negligibly small.

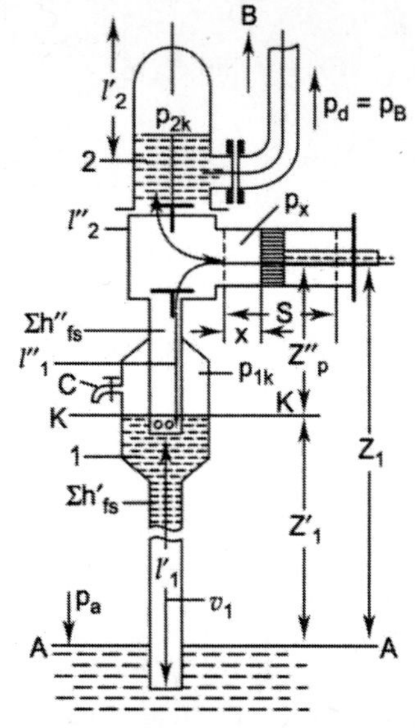

Figure 1.8. Working of reciprocating pump with air vessels

1.6.3 Delivery

The air vessel 2, connected very close to the pump, receives the liquid from the pump under unsteady non-uniform flow condition in the cylinder of length x and the delivery pipe length between cylinder and air vessel l_2''. The pressure p_{2k} is smaller than the delivery pressure in the cylinder p_x and larger than the delivery pressure p_d (p_B). The air and the liquid from the cylinder pass through the delivery valve to the delivery air vessel. The liquid level and the pressure in the air vessel goes on changing due to release of air from the liquid. An unsteady flow exists in the pipe length l_2''. At the same time, uniform flow takes place from air vessel to the delivery point through the delivery pipe l_2'. Writing down the equation

$$\frac{p_x}{\gamma} = \frac{p_d}{\gamma} + \frac{v_d^2}{2g} + Z_2 + \sum h_{fd} + \frac{l_2'' + x}{g}\,\omega^2 r \cos\phi \qquad \text{...(1.48)}$$

where l_2'' is the delivery length between cylinder and air vessel. Usually this air accumulated in delivery air vessel, is removed automatically through the safety valve C (figure 1.8). Normally in delivery air vessels liquid is permitted to fill up to 2/3 of its total volume. The volume of the air vessel is determined by the permitted unevenness in pressure δ_p inside air vessel.

$$\delta_p = 2\,\frac{p_{max} - p_{min}}{p_{max} - p_{min}} \approx 0.01 \text{ to } 0.05 \qquad \text{...(1.49)}$$

This is quite sufficient for practical purposes. This unevenness in pressure can be attained with 27FS for single pump, 10FS for two pumps, 0.5 FS for three pumps and 0 2FS for 4 pumps.

1.7 SUCTION CONDITION AND ALLOWABLE SPEED

The design, commercial exploitation and regulation of any reciprocating pump purely depend upon its suction conditions. The pump speed is always determined, from the maximum allowable suction lift. Flow separation takes place, when the pressure p_x, (figure 1.8) inside the cylinder falls below the vapour pressure (p_{vp}) of the liquid at the pumping temperature. Minimum pressure p_{min} occurs at three positions of the piston, (*a*) at the beginning of the suction stroke, where the inertia head is maximum. The total suction head will be $h_s + h_{as}$ (frictional head is zero), (*b*) at the middle of the suction stroke, where the friction losses are maximum, and the total suction head is $h_s + h_{fs}$ and (*c*) at the end of the delivery stroke, where the inertia head is high (*i.e.*,) $h_d - h_{ad}$. In all these cases, the cylinder space is filled with gas which occupies larger volume. During delivery stroke, the gas is pumped out along with the liquid. At higher pressures, the vapour condenses into liquid, which occupies very small volume, as a result of which, flow separation takes place and the volume rate of flow is reduced. Taking suction sump pressure is equal to atmospheric pressure and neglecting the velocity head, since it is zero, (Figure 1.8)

$$\frac{p_x}{\gamma} = \frac{p_{at}}{\gamma} - h_s - \sum h_{fs} \qquad \text{...(1.50)}$$

where $\sum h_{fs} = h_{ac} + h_{f(p)} + h_{f(v)}$

In suction pipe $h_{ac} = \frac{l_1 + x}{g}\omega^2 r \cos\phi$ is the inertia head, with out air vessel, and $= \frac{l_1'' + x}{g}\omega^2 r\cos\phi$, is the inertia head with air vessel $h_{fp} = \frac{fl_1(\omega r \sin\theta)^2}{2gd}$ is the loss of head due to friction when there is no air vessel and $= \frac{fl_1''(\omega r\sin\theta)^2}{2gd}$ is the loss of head due to friction, before the air vessel due to non uniform flow for the length l_1'' and $+ \frac{fl_1' v_s^2}{2gd_s}$ is the loss of head due to friction due to uniform flow in the pipe length l_1' after the air vessel, when there is air vessel and $h_{f(v)} = K\frac{(\omega r\sin\theta)^2}{2gd}$ is the loss of head in the suction valve.

Considering the losses, $\sum h_f$, in detail, at the beginning of the suction stroke when $\phi = 0°$, $\frac{l_1 + x}{g}$, $\omega^2 r\cos\phi$, the inertia head is maximum when there is no air vessel, $\frac{l_1'' + x}{g}$ $\omega^2 r\cos\phi$, the inertia head is maximum when there is air vessel. $\frac{fl_1(\omega r\sin\theta)^2}{2gd_s}$ is the frictional loss of head in the suction pipe when the air vessel is not fitted and is maximum at the middle of the stroke when $\phi = 90°$. $h_{f(v)}$ is the suction valve loss = $\frac{K(\omega r \sin\theta)^2}{2g}$ and is maximum at the middle of the suction stroke when $\phi = 90°$. Substituting these valves, in the equation (1.30), it is found that cavitations can occur at two positions of the suction stroke. (1) At the beginning of the suction stroke due to high inertia head *i.e.,*

$$h_{s(cr)} = \frac{p_{at} - p_{vp}}{\gamma} - \left[K\frac{(\omega r)^2}{2g} + \left(\frac{l_1 + x}{g}\omega^2 r\right)\right] \quad ...(1.51)$$

and (2) at the middle of the suction stroke due to high friction losses (*i.e.,*)

$$h_{s(cr)} = \frac{p_{at} - p_{vp}}{\gamma} - \left[\frac{fl_1(\omega r)^2}{2gd} + \frac{K(\omega r)^2}{2g}\right] \quad ...(1.52)$$

The allowable suction head should be h_s (allowable) = $h_{s(cr)}$ – (1 to 1.5 m)

In the same manner, considering the delivery side the maximum pressure can occur at the end of the delivery stroke *i.e.,*

$$h_{d(cr)} = h_{d(T)} = \frac{p_{at} - p_{vp}}{\gamma} - \left[K\frac{(\omega r)^2}{2g} + \left(\frac{l_2 + x}{g}\omega^2 r\right)\right] \quad ...(1.53)$$

Critical suction head will be when $p_x = p_{vp}$

$$h_{s(cr)} = \frac{p_{at}}{\gamma} - \frac{p_{vp}}{\gamma} - \sum h_{fs} \quad ...(1.54)$$

Allowable suction head is taken a little above to avoid cavitation. Allowing 1to1.5 m as allowance allowable $h_s = h_{s(cr)}$ – (1 to 1.5). Corresponding allowable delivery head will be h_d (allowable) = $h_{d(cr)}$ – (1 to 1.5 m). The maximum value of h_{cr} is taken to determine the speed.

The allowable speed $n = n_{max}$ is determined from the value of ω from the equation where the value of h_{cr}, calculated, is maximum.

$$n_{max} = \frac{30}{\pi}\sqrt{\frac{g}{l_1'' r}\left(\frac{p_{at} - p_{vp}}{\gamma} - h_{cr} - h_{v\,max}\right)} \quad ...(1.55)$$

The speed of the reciprocating pump is normally very small and varies between 150 rpm to 350 rpm. Figure 1.9 (*a*) gives a typical cavitation test results for a reciprocating pump. Figure 1.9 (*b*) gives the values of volumetric efficiency as a function of suction head, speed and the nature of pumping liquid. Figure 1.9 (*c*) gives the values of h_s allowable for different viscous liquids with the boundary conditions for speed selection for reciprocating pumps. Figure 1.9 (*a*), 1.9 (*b*) give the values of h_s allowable for different viscous liquids with the boundary conditions for speed selection for reciprocating pumps. /2,4,8, 14./

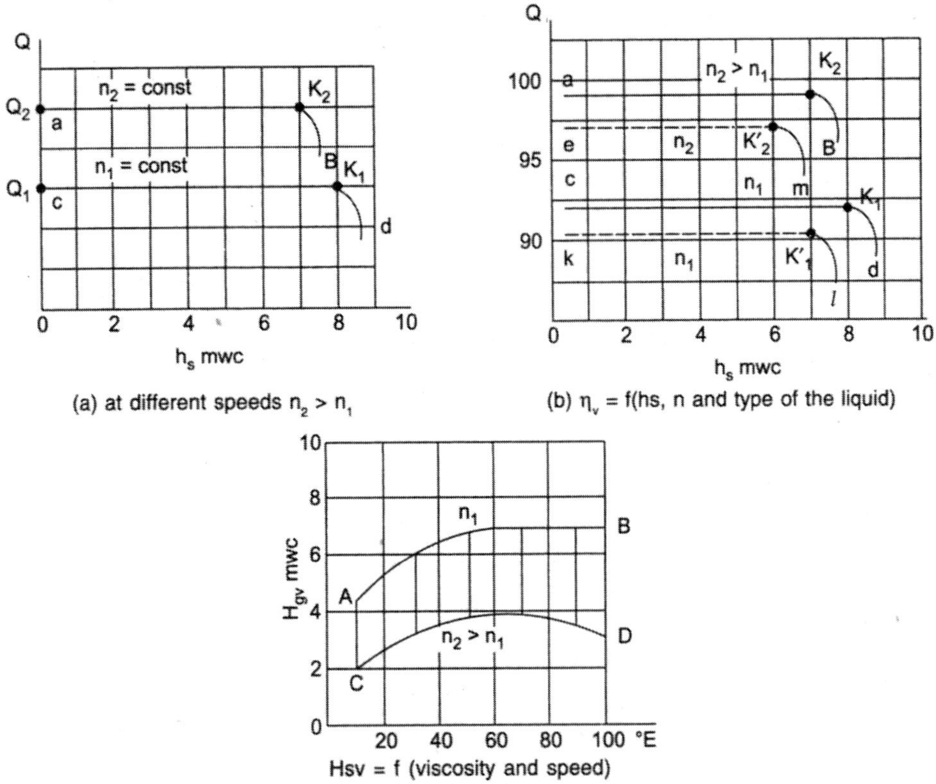

Figure 1.9. Cavitational characteristics of reciprocating pump

1.8 LOSSES AND EFFICIENCY CALCULATIONS

Hydraulic, volumetric and mechanical losses and corresponding efficiencies are determined to obtain overall efficiency.

1.8.1 Volumetric Losses ΔQ and Volumetric Efficiency η_v

Leakage losses through the clearance between the piston and the cylinder, through the clearance between the valve and valve seat, through the stuffing box as well as loss of flow due to non filling of the space between the piston and the cylinder during suction stroke, due to the change in the velocity of the fluid entering with respect to the velocity the piston, due to gas presence in the form of bubbles, due to high speed of the pump and subsequent flow separation, due to delay in opening of suction and delivery valves, which results in reduced flow, from pipe to cylinder, due to air entry into the cylinder through stuffing box are the main reasons for the reduction in the discharge. Total volumetric loss will be $\Delta Q_l + \Delta Q_v$, where ΔQ_l – loss due to leakage and ΔQ_v is the loss due to insufficient quantity filling of liquid inside the cylinder.

Actual volume $\quad Q_a = Q_{TH} - \Delta Q_l - \Delta Q_v = Q_{TH} - \Delta Q$, where $\Delta Q = \Delta Q_l + \Delta Q_v$

and volumetric efficiency, η_v

$$= \frac{Q_a}{Q_{TH}} = \frac{Q_{TH} - \Delta Q}{Q_{TH}} = \frac{Q_a}{Q_a + \Delta Q} = 1 - \frac{\Delta Q}{Q_{TH}} \quad ...(1.56)$$

Volumetric efficiency in reciprocating pumps vary from $\eta_v = 0.8$ to 0.9 and is always determined by experiments. One such result is given in figure 1.10, curve AB is for low head high discharge and curve CD is for high head low discharge pumps. Curve BM and DN correspond to the pump operation very near to cavitation. /2,4/

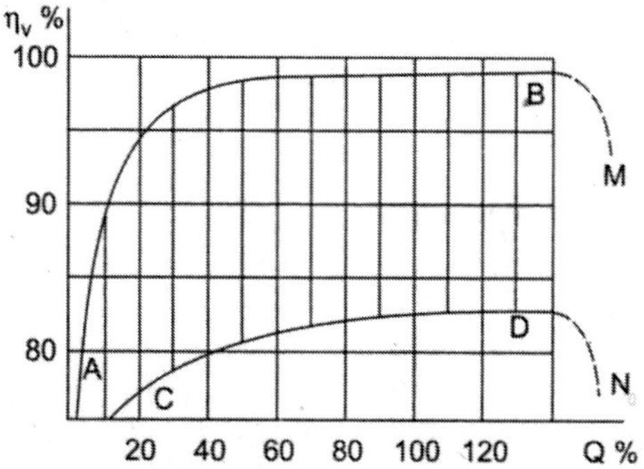

Figure 1.10. $\eta_v = f(\%Q)$

1.8.2 Hydraulic Losses (ΔH) and Hydraulic Efficiency (η_h)

Hydraulic losses, ΔH, in reciprocating pumps, mostly attribute to losses in valves, due to sudden contraction and sudden expansion in flow, high velocity of flow through valve passages and pipe frictional losses, major and minor losses.

Hydraulic efficiency $$\eta_h = \frac{H_a}{H_{Th}} = \frac{H_{Th} - \Delta H}{H_{Th}} = \frac{H_a}{H_a + \Delta H} = 1 - \frac{\Delta H}{H_{TH}} \quad ...(1.57)$$

Hydraulic losses are negligible in reciprocating pumps when compared to volumetric and mechanical losses. Normal value of η is 0.97 to 0.99

1.8.3 Mechanical Losses ΔN and Mechanical Efficiency (η_m)

Mechanical losses depend upon the type of pump construction and method of lubrication adopted. Mechanical losses attribute to frictional losses between piston and cylinder, stuffing box losses, bearing losses and other, crank pin losses, crankshaft bearing losses etc. The difference between the power input to the crank shaft or the power output from the prime mover N and the power available of the piston $\left(N_{ip} = \frac{\gamma Q_{th} H_{th}}{\text{const K}}\right)$ is the total mechanical losses ΔN.

$$N - \Delta N = N_{ip} = \frac{\gamma Q_{th} H_{th}}{\text{const K}}$$

Mechanical efficiency of the reciprocating of the pump

$$\eta_m = \frac{N_{ip}}{N} = 1 - \frac{\Delta N}{N} = \frac{N_{ip}}{N_{ip} + \Delta N} = \frac{N - \Delta N}{N} \quad ...(1.58)$$

Mechanical efficiency η_m in reciprocating pumps vary as varies from 0.85 to 0.96

Overall efficiency $\eta = \eta_v \eta_n \eta_m$ ranges from 0.5 to 0.65.

Figure 1.11 gives an approximate value in overall efficiency. Efficiency curve 0A belongs to medium head high discharge pumps (Q = 100 to 300 m^3/hr and H = 40 to 60 m) and curve OB belongs to medium discharge and high head pumps (Q = 25 to 80 m^3/hr and H = 100 to 500 m) /2,4/

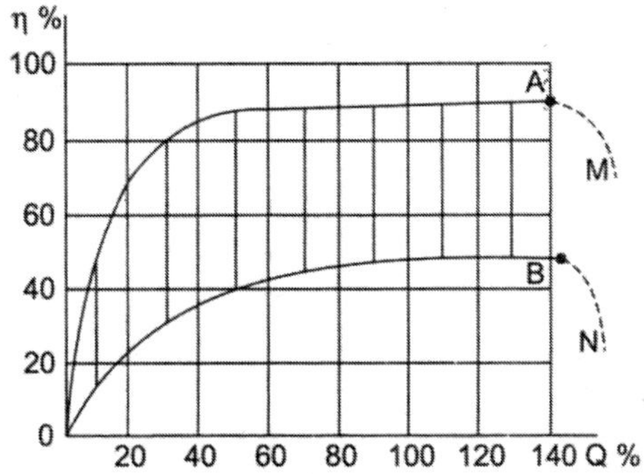

Figure 1.11. $\eta_m = f(Q\%)$

In all positive displacement pumps efficiency remains almost constant for a wide range of flow rate but during cavitation, efficiency drops down rapidly, curves AM and BN.

1.8.4 Work done, Power and Efficiency of Reciprocating Pump

Work done by the reciprocating pump can be written as W.D = $k_q (p_d - p_s)$ A.S + T, where A and S are the area and stroke length of the cylinder, p_d and p_s are the average pressures at delivery and at suction, T is the mechanical losses in stuffing box, bearings and at connecting rod joints T = (0.03 to 0.10) $k_q (p_d - p_s)$ A.S

Power consumption $$N_i = \frac{(W.D).n}{60 \times const} \text{ kW} \qquad ...(1.59)$$

where n is the speed in rpm.

1.9 PERFORMANCE OF RECIPROCATING PUMPS

Performance of a reciprocating pump is represented by the variation of flow rate (Q), efficiency (η) and power (N) as function of total head (H). However performance depends upon the type of pump construction, exploitation consideration etc., most important performance curve is the performance, drawn in H = f(Q) form. Figure 1.12 indicates the performance in H-Q form. Lines AB ,CD MK, and EF are theoretical H-Q curve. Since head is independent of quantity of flow in all positive displacement pumps, the curves are straight lines ; lines MK and EF, parallel to X axis, are obtained by throttling the delivery valve for head, where as, lines AB and CD, parallel to Y – axis for quantity, are obtained by changing the speed.

However actual H-Q curve will be CR. The reduction in quantity is due to higher leakage flow when total head increases. The horizontal difference between curves CD and CR represents ΔQ_l, the leakage flow. Sometimes, at high pressures, due to improper closing and opening of valves, the flow reduces considerably, especially, at higher delivery pressures, or under high vacuum. H-Q curve, changes to AGL instead of CR. The volumetric loss will be more in these cases.

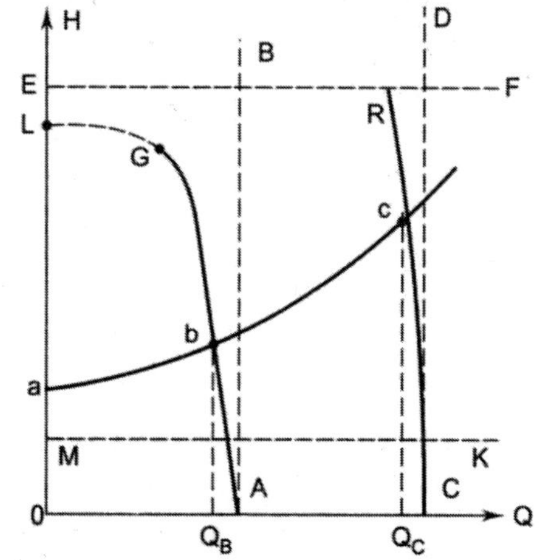

Figure 1.12. H = f(Q) graph for a reciprocating pump

1.10 REGULATION OF PUMPS

Regulation in positive displacement pumps is always carried out for flow rate and total head. Flow rate is changed by changing the speed, and total head is changed by the control valve at the delivery pipe. But total head automatically changes, to accommodate friction in the pipe. In all these cases, the power changes proportionately. In figure 1.12, the curve *ac* indicates the system curve. In order to change the operating point from *b* to *c*, the speed must be increased. Correspondingly the power is increased. Normally a safety or relief valve is provided at the delivery pipe. This valve opens automatically, when the delivery pressure reaches the safe operating pressure of the pump. The liquid is released and the pressure drops down below the upper limit of operation. In case the pumping liquid is other than water such as viscous liquid, the released liquid from the safety or relief valve is brought back to the suction pump. A valve is fitted between delivery pipe and suction sump. However material strength of the pump must be considered while the total head is increased. Cavitation must be considered, when the pump operates at very high flow rate or at very low head.

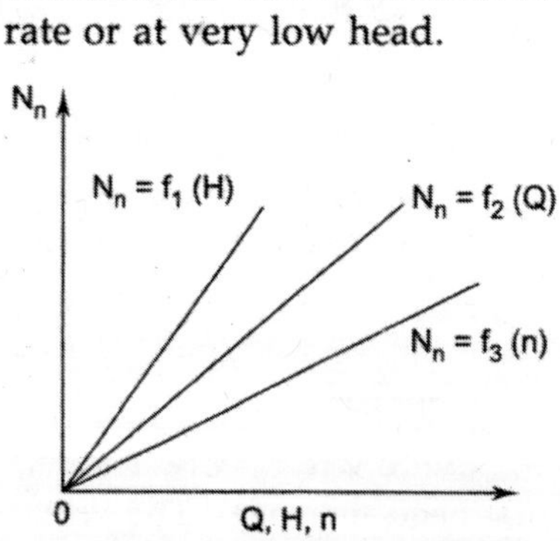

Figure 1.13. Variation of N with respect to Q, H, n,

1.11 DETERMINATION OF MAJOR DIMENSIONS OF PUMP, SUCTION AND DELIVERY PIPE DIAMETERS

Two types of calculations must be considered while determining the pump sizes: (1) Hydraulic and (2) Mechanical. For any design, the available parameters are flow rate Q, total head H, delivery pressure p_2 or suction conditions (or) suction lift h_s and liquid property such as viscosity density, temperature.

1.11.1 Number of Pumps and Types of Pumps

Increase in the number of pumps as well as double acting pumps reduce the size of the pump. However valves cannot be reduced in dimension, similar to the reduction of piston diameter due to cavitation, as well as flow separation at the valve. Cavitation, can be reduced, by increasing the suction pipe diameter. Recommended conditions are 20 to 120 double acting in a minute for smaller flow and 100 to 200 double acting/min for higher flow rate.

1.11.2 Selection of Cylinder Diameter D

For design purpose, the diameters of piston and cylinder are considered equal although there is a sliding clearance between piston and cylinder. Since this clearance is very small it is neglected.

If there are Z number of pumps, operated in parallel in one crank shaft and Q_{TO} is the total quantity, the quantity per pump $Q = (Q_{TO}/Z)$.

$\frac{Q}{\eta_v} = z\frac{\pi D^2}{4}Sn$, where z = 1 for single acting and 2 for double acting. The piston rod diameter is neglected for double acting units.

$$D = \sqrt{\frac{4Q}{\pi z S n \eta_v}} \qquad ...(1.60)$$

Theoretical Quantity, Q_{TH},

for single acting $$Q_{TH} = A\frac{Sn}{60} = \frac{\pi D^2}{4}\frac{Sn}{60}\ m^3/\text{sec} \qquad ...(1.61)$$

for double acting $$Q_{TH} = \frac{Sn}{60}\left(A + (A - a)\right) = (2A - a)\frac{Sn}{60} = \left(1 - \frac{a}{2A}\right)2A\frac{Sn}{60} = \phi.\frac{\pi D^2}{4}\frac{Sn}{60} \qquad ...(1.62)$$

Where $\phi = 2\left(1 - \frac{a}{2A}\right)$ for double acting and $\phi = 1$ for single acting

a – Area of the piston rod

A – Area of the piston

D – Diameter of the piston

d – Diameter of the piston rod

S – Stroke length

n – Speed in rpm

The reduction in the quantity of flow is attributed to the following reasons:

1. Due to the delay, in opening and closing of the valves, a reversed flow is created, resulting in a reduction in the flow into cylinder during the suction stroke.
2. Due to sudden change in the area between cylinder and valve gap, the cylinder can not be filled to its full volume within the time of suction valve opening and closing. The fluid flow velocity, through the valve becomes very high; as a result, flow separation takes place near the piston surface during suction stroke.
3. Leakage through the valve due to improper seating of the valves, leakage through the stuffing box, leakage through the clearance between piston and the cylinder.
4. Due to filling up of vapour, inside the cylinder, due to reduced pressure, below vapour pressure

Most influencing factor is the reduction of flow rate due to the presence of gases and water vapour inside the cylinder space, evolving due to very low pressure, below vapour pressure, entering through the gap between valve and valve seat due to improper valve seating as well as through the gap between piston and cylinder due to pressure difference prevailing between both sides of the piston. Actual flow rate takes place only when the pressure inside the cylinder is equal to the prevailing suction pressure in the pipe.

The reduced flow rate can be subdivided into three categories:

1. One without energy loss, example: presence of air inside the cylinder (work of compression equals work of expansion),
2. The second with energy loss, example: delayed action of opening and closing of valves, which increases hydraulic loss only to a smaller extent and
3. The third the leakage flow through the clearances, which increases the hydraulic loss considerably.

The total reduction in flow rate 'q' can be written as $q = q_1 + q_2$ where q_1 is the flow reduction without energy loss and q_2 is the flow reduction with energy loss $DN = \gamma q_2 H_{th}$ where γ is the specific weight of the liquid in N/m^3 and H_{th} is the theoretical head.

Taking

The flow coefficient $$\lambda = \frac{Q}{Q_{th}}, \qquad \text{...(1.63)}$$

The volumetric efficiency η_v

$$\eta_v = \frac{Q}{Q + q_2} \qquad \text{...(1.64)}$$

Coefficient of cylinder filling $$K_v = \frac{Q + q_2}{Q_{th}} = \frac{Q_{th} - q_1}{Q_{th}}$$

Hence $$\eta_v . K_v = \frac{Q}{Q+q_2} . \frac{Q+q_2}{Q_{th}} = \frac{Q}{Q_{th}} = \lambda \quad ...(1.65)$$

The flow coefficient λ fluctuates, depending upon the manufacturing processes adopted, pump size, viscosity of the liquid, number of double strokes made in a minute by the pump, *i.e.*, No. of revolutions/minute, the delivery pressure and actual operating conditions. Experimental valves of λ are given in Table 1.1.

Table 1.1

D < 50 mm	λ = 0.85 to 0.90
D = 50 to 75 mm	λ = 0.90 to 0.95
D = 75 to 100 mm	λ = 0.95 to 0.98
D > 100 mm	λ = 0.98 to 0.99
For double acting	λ = 0.85 to 0.95
For air	λ = 0.70 to 0.80

λ remains same for viscous liquid as well as for water pumping for normal suction lift of 3.5 to 4 m. For higher suction lifts λ reduces due to the reduction of k_v.

Taking m = S/D, where S is the stroke length,

The flow rate

$$Q = \phi\lambda \frac{\pi D^2}{4} S. \frac{n}{60} = \phi\lambda m \frac{\pi D^3}{4} \frac{n}{60} \text{(or)} D = \sqrt[3]{\frac{4 * 60 * Q}{\pi\lambda\phi m n}} \; (m) \quad ...(1.66)$$

In most of the designs the average velocity u_{av} is considered as a reference velocity and is given as

$$u_{av} = \frac{2.S.n}{60}$$

In such cases, the flow rate

$$Q = \phi\lambda. \frac{\pi D^2}{4} \frac{u_{av}}{2} m^3/\text{sec} \text{(or)} D = \sqrt{\frac{8Q}{\phi\lambda\pi u_{av}}} \quad ...(1.67)$$

If Z is the number of pumps, operated in one crankshaft

$$Q = Z.\phi\lambda. \frac{\pi D^2}{4} \frac{u_{av}}{2}$$

the diameter $$D = \sqrt{\frac{8Q}{Z\phi\lambda\pi u_{av}}} \quad ...(1.68)$$

Experimental results for 'u_{av}' are given in figure 1.14 (*a*) as a function of stroke length 'S'.

u_{av} reduces when viscosity of the liquid increases. Figure 1.14(*b*) gives the values of 'U_{av}' for different values of viscosity of the liquid (in °E) /2,22,23/

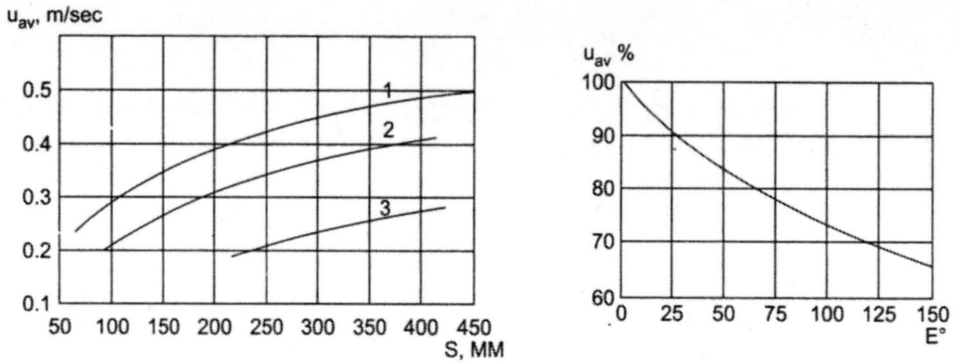

(a) Graph $u_{av} = f(S)$ (b) Graph $u_{av} = f(°E)$

Figure 1.14.

Speed of the pump varies depending upon the number of pumps connected to a single crankshaft. For 4 pumps n = 40 to 120 rev/minute, for two pumps n = 20 to 30 rev/min and for single acting n = 15 to 30/min.

1.11.3 Selection of (S/D) Ratio 22, 23, 24

For better operation and force considerations S/D is selected between 0.75 and 2.8. For transport applications S/D lies between 1.0 to 1.5. The lower range of S/D in both cases attribute to low head and high specific speed units and higher range is for high head and low specific speed units. Reduction of S/D reduces the force acting on the connecting rod, which is advantageous for high head units.

The value m(S/D ratio) depends upon the pump type. For plunger pumps m = 1 to 1.2. But increase in diameter increases the size of the pump and increase in stroke length increases the acceleration, which in turn, increases the load on the pump.

The value of ϕ ranges from 0.98 to 0.99 in larger pumps and almost 100% in smaller pumps. However, in high pressure, low and medium discharge pump it ranges from 0.85 to 0.87.

1.11.4 Selection of Pipe Size

The pipe diameter,
$$d = \sqrt{\frac{4Q}{\pi C}} \quad ...(1.69)$$

The velocity 'C' for water will be, C_1 (C_s) = 0.75 to 2 m/s at suction C_2 (C_d) = 1 to 3 m/s at delivery. For oil and for low viscous fluid C_1 = 0.5 to 1.5 m/s. C_2 = 1 to 2 m/s.

Frictional losses reduce when diameter of the pipe is increased.

Diameter
$$d_s = \sqrt{\frac{4Q}{\pi C_s}} \text{ and diameter } d_d = \sqrt{\frac{4Q}{\pi C_d}} \quad ...(1.70)$$

1.12 EXPLOITATION OF PUMPS

Before starting any new pump the following must be checked:

1. Foundation must be checked. Pump must be firmly fixed to the foundation.

2. Operation of the valves must be checked.
3. Proper lubrication must be provided.
4. The suction pipe should be filled with the pumping liquid and all valves in the pipe must be kept in full open condition.

While running:

(*a*) The control and measuring instruments must be in good condition.

(*b*) Proper and adequate capacity meters must be used.

(*c*) Stuffing box must be properly maintained.

(*d*) Overheating of pump must be avoided.

(*e*) Operating time and other conditions must be carried out as per the manufacturer's instruction manual.

(*f*) Stopping procedure must be strictly followed as per the recommendations. After stopping any problems noticed during running of pump must be attended immediately.

The following irregularities must be avoided totally:

1. Valves in pipes must not be in closed condition under idle conditions, before starting and after stopping the pump.
2. To avoid cavitation, the liquid level in the suction sump should not be lower than the recommended value even while running.
3. Entry of air due to improper tightening of joints.
4. Unacceptable noise and vibration when pump in operation.

CHAPTER **2**

RADIAL PISTON PUMPS

2.1 INTRODUCTION

Radial piston pumps and axial piston pumps are used for very high pressures and for very small flow rates. The advantages of these pumps are

1. Compact, occupies smaller space, when compared to other pumps.
2. Can change the flow rate without altering the speed and delivery pressure, by altering the stroke length.
3. Consistently the pressure and flow rate are constant.
4. Higher efficiency at high pressures and remains constant for a wide range of flow rate.
5. Flow can be reversed by changing the direction of rotation.

Rotary piston pumps are classified into two groups;

1. Radial piston pumps and
2. Axial piston pumps.

These pumps can be operated as fluid pumps, as well as, fluid motors. They are used in airplanes, rockets, satellites, ships and other transport vehicles, dealing with hydraulic circuits such as lifting and other operations. These pumps are operated for a flow rate of 0.5 to 25 m^3/hour and for a delivery pressure of 100 to 200 ata. Speed varies from 500 to 1000 rpm. Due to high radial force developed at high pressures and at high discharge operating conditions, the size of the bearing and the size of the flow passage, become very large, which increases the size of the pump. That's why, these pumps are used only for very low flow rates.

2.2 PRINCIPLE AND CONSTRUCTION OF RADIAL PISTON PUMPS

The constructional details of a radial piston pump are given in figure 2.1. A rotor 5 is located, inside a stationary pump casing 1, with horizontal axis *x-x* and vertical axis *y-y*.

The rotor has a number of circumferentially equi-spaced and radially located cylindrical holes, into which, a number of plungers 8 are installed (Fig. 2.1).

On one side, these plungers are connected, with a ball and socket arrangement or with rollers or with ball bearings, to a base 9, which is in the form of an arc of a circle. The base is always in contact with a moving outer ring 2. This ring can be moved axially along the direction *x-x* and *z-z* with the help of screw mechanism 3, either towards left or towards right, along the axis *x-x*, by which, the eccentricity e and the stroke of plungers can be changed. The rotor contains a non moving cylindrical hollow space 4 at the centre. This hollow space is diametrically divided into two semicircular portions, the top space 7 is the suction and the bottom portion 6 is the delivery. These two spaces are connected to the corresponding suction and delivery pipes respectively. When the ring 2 is displaced to the right hand side, the axis *z-z* will be at the right side, with respect to axis *y-y* and vice versa. The distance between the two axis *y-y* and *z-z* is the eccentricity '*e*'. The ring can be moved either to the left hand side or to the right hand side by adjusting the screw, there-by the eccentricity '*e*' can be brought either to the right side or to the left side. The magnitude of the eccentricity can also be controlled by which the stroke length of the plunger, sliding inside the cylinder, can be adjusted, which in turn changes the flow rate. When the eccentricity is at the right side, *i.e.*, when the ring is moved toward the right side, the rotor rotates along with the plungers in the anticlockwise direction as indicated by the arrow. Simultaneously, the plunger reciprocates inside the cylinder in the radial direction. The stroke of each plunger increases, when the cylinder is located below *x-x*, *i.e.*, at the space 7. During the rotation of the rotor, the plunger, at this area, creates a vacuum inside the cylinder. The liquid, coming from the suction pipe occupies the space 7, and then goes into the cylinder *i.e.*, suction stroke takes place. When the plunger reaches the upper half, the area 6, the plunger stroke is in the downward direction. The fluid comes out of the cylinder and occupies the space 6 and then goes out to the delivery pipe.

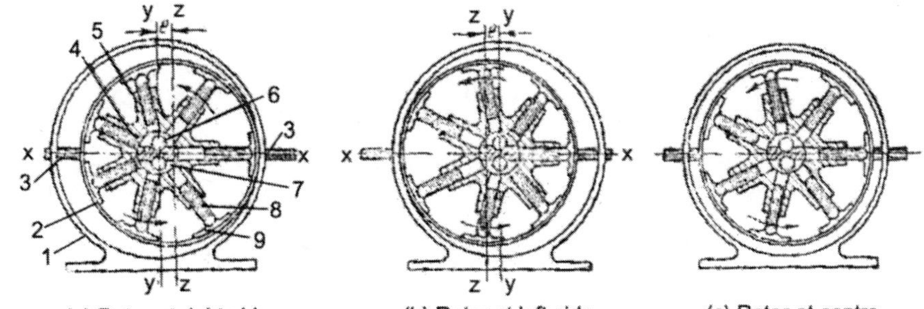

(a) Rotor at right side (b) Rotor at left side (c) Rotor at centre

Figure 2.1. Schematic diagram of a radial piston pump

Each plunger makes one complete up and down reciprocating motion during one revolution of the rotor. The liquid is sucked, when the plunger comes to the space 7 and is delivered, when the plunger comes to the space 6, during the rotation.

If the eccentricity '*e*' is in the left side the suction and delivery will be interchanged. The direction of liquid flow is reversed, *i.e.*, space 7 become delivery and space 6 becomes suction. If the eccentricity is zero, *i.e.*, if the axis *y*-*y* and z-z coincide, there will not be any flow even though the rotor rotates. The plungers will not have any reciprocating motion. Hence the eccentricity '*e*' provides a facility to vary the flow rate. Depending upon the quantity required, the number of plungers Z vary from 5 to 9. In radial piston pumps, however, more uniform flow is achieved with more number of plungers. Also flow rate can be increased by having a number of rotors in one shaft, *i.e.*, a number of pumps in one shaft with a common delivery and suction as similar to the operation of pumps in parallel.

The main advantages of these pumps are

1. Flow rate can be altered by changing the eccentricity and
2. The flow can be reversed in the pump *i.e.*, suction and delivery ports can be interchanged by bringing the eccentricity to the opposite side.

Normal working of radial piston pumps can be attained, only when the plunger is continuously and firmly in contact with the moving guide ring at all times of operation of the pump. This is essentially very important, especially, when the plunger moves radially inwards, during delivery stroke, since removal of plunger from the moving ring may take place. This occurs not only in radial piston pumps but also in axial piston pumps. When the plunger moves towards the guide ring, a compressive force is created on the guide ring, due to centrifugal force created by the mass of the plunger. At the same time the liquid pressure, from the suction chamber acts on the other side of the plunger. The resultant force acts on the guide ring.

Consistency in pump operation can be attained at constant speed of rotation. At relatively lower speeds of 550 to 1000 rpm, and at quite often changing the eccentricity conditions, which is required in some of the installations such as in ships, it is essential to have perfect contact between the plunger and the guide ring. This is achieved by ball and socket arrangement or with rollers fitted, at outer end of the plunger, both these give a constant contact with the inner side of the guide ring. Nowadays, ball bearings are used instead of rollers, which reduces the friction between the ring and the plunger.

2.3 EQUATION OF MOTION AND FLOW RATE CALCULATIONS

Referring to figure 2.2, by considering its Kinematic motion, and by using cosine theorem, the movement of the plunger can be written as

$$AO_1^2 = AO^2 + OO_1^2 - 2AO \times OO_1 \cos(180 - \phi) \quad \text{...(2.1)}$$

where

O	is the rotor centre or shaft axis
O_1	is the centre of the guide ring
R	is the radius of the guide ring
r	is the rotor radius
e	is the eccentricity
S = AB	is the stroke length of the plunger
ϕ	is the angle of rotation
ω	is the angular velocity,

Referring to figure 2.2

$$AO_1 = R = r + e, AO = r + S, OO_1 = e, \cos(180 - \phi) = -\cos\phi.$$

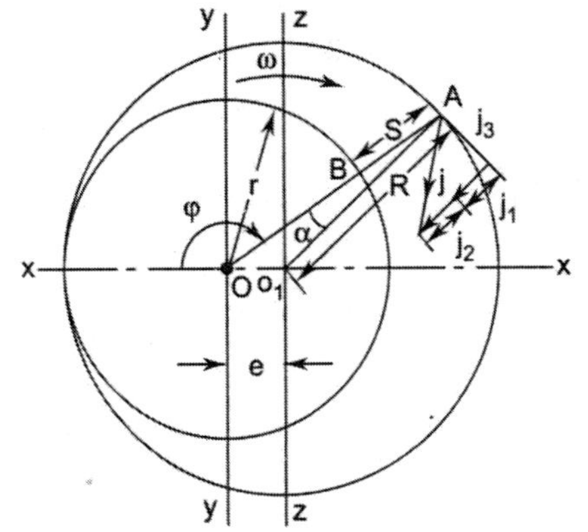

Figure 2.2. Kinematic diagram for a radial piston pump

Substituting these values in equation 2.1,

$$(r+e)^2 = (r+S)^2 + e^2 + 2e(r+S)\cos\phi$$

Solving for the stroke S

$$S^2 + 2S(r + e\cos\phi) + 2er(\cos\phi - 1) = 0$$

$$S = -(r + e\cos\phi) + \sqrt{(r + e\cos\phi)^2 - 2er(\cos\phi - 1)}$$

$$= -(r + e\cos\phi) + \sqrt{r^2 + e^2\cos^2\phi + 2er} \qquad ...(2.2)$$

Maximum stroke is attained when $S_{max} = 2e$ at $\phi = 180°$. If a graph $s = f(\phi)$ is drawn, the curve will be a sinusoidal curve with S_{max} at $\phi = 180$. $S = 0$ when $e = 0$. The flow rate and the motion of the plunger is determined from the relative velocity of the plunger, w_1. The relative velocity of the plunger w_1 is calculated as

$$w_1 = \frac{ds}{dt} = e\sin\phi\frac{d\phi}{dt} - \frac{e^2\cos\phi\sin\phi\dfrac{d\phi}{dt}}{\sqrt{r^2 + e^2\cos^2\phi + 2er}} \qquad ...(2.3)$$

Since $\dfrac{d\phi}{dt} = \omega = \dfrac{2\pi n}{60}$ the angular velocity of the rotor, equation 2.3 is modified and rewritten as

$$w_1 = \frac{2\pi ne}{60}\left(1 - \frac{e\cos\phi}{r}\right)\sin\phi \qquad ...(2.4)$$

It is seen, from the equation (2.4), that the velocity w_1 is not constant, instead, varies in sinusoidal form as similar to that in reciprocating pumps *i.e.*, the flow is unsteady. In figure 2.3, the change in velocity w_1 with respect to the angle ϕ, is shown. $w_1 = 0$ when $\phi = 0°$ and $180°$ and w_1 is maximum when $\phi = 90$. The theoretical flow rate for one plunger q_{th} is

$$q_{th} = w_1 A = \frac{2\pi n.e.A}{60}\left(1 - \frac{e\cos\phi}{r}\right)\sin\phi \qquad ...(2.5)$$

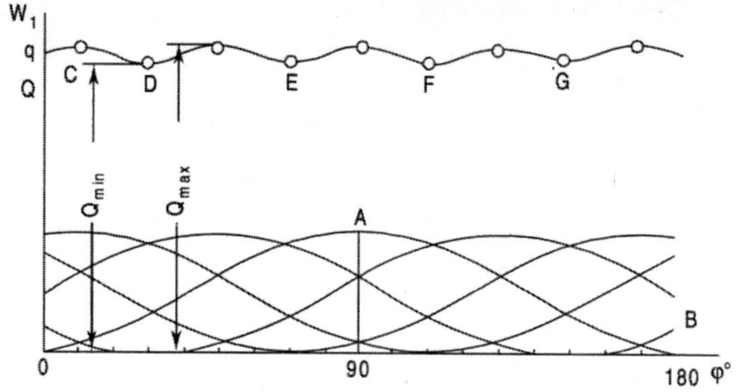

Figure 2.3. Change of Flow rate and plunger velocity with respect to angle of rotation of rotor

Equation 2.5 is represented by the curve OAB in figure 2.3. Total quantity (Q_{th}) delivered by 5 plungers is given in this figure, which indicates a very little fluctuation in flow rate. Thus fluctuation can be reduced by having more number of plungers.

Total average flow rate

$$Q_{th(av)} = q_{th\,(av)}\,Z = \frac{\pi d^2}{4}.2enZ = 0.5\pi d^2\,neZ \qquad ...(2.6)$$

where Z is the number of cylinders or plungers and d is the plunger diameter.

2.4 FORCES ACTING ON THE RADIAL PISTON PUMP

Three forces act on the radial piston pump namely

1. Forces due to the fluid pressure (figure 2.4) acting on the plunger and the cylinder.
2. Inertia force arising out, due to the rotary and reciprocating motion of plunger and its parts.
3. Frictional force between the plunger and the cylinder, as well as between the base and the guide ring.

Determination of wall thickness for the pump components is based on the maximum force, acting on the pump. Graphical method will be easier and quicker, since an approximate value is sufficient for the calculation of thickness. However a detailed analyses and calculation for the determination of these forces will be carried out at a later stage.

Based on the predetermined dimensions for the pump and other connecting dimensions for the flow rate required, the inner radius of the guiding ring R is determined. Referring to the figure 2.4, the force acting on one cylinder, due to delivery pressure depends upon the frictional losses in delivery pipe. Maximum unit pressure acting on the plunger, p_d, occurs at the maximum angle of α_{max}, which can be written as

$$p_d = p_{max} + \sum \gamma\, h_f$$

where p_{max} is the maximum pressure on the cylinder and $\sum \gamma h_f$ is the frictional losses in delivery pipe. Maximum force due to fluid pressure is

$$P = p_d \frac{\pi d^2}{4}$$

Under eccentric condition of the shaft axis O with respect to the guiding ring O_1, the force P is resolved into two components, namely force N, directed along the radius of the ring and force T, acting perpendicular to the axis of the plunger.

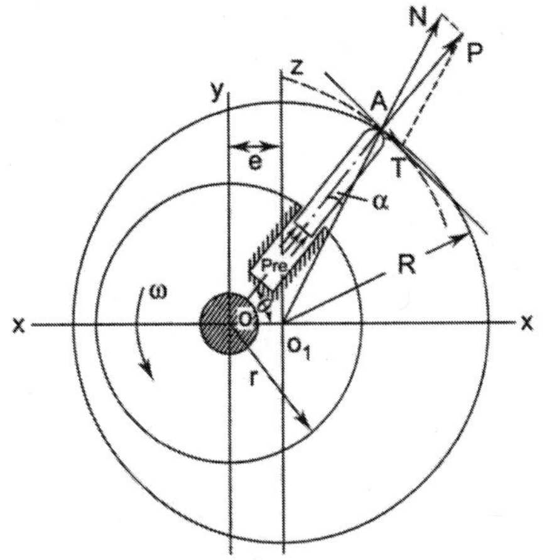

Figure 2.4. Hydraulic forces acting on the plunger

The force N gives a compressive force on the guide ring, and force T acting perpendicular to the plunger axis, gives a moment of friction, which acts on the shaft and to the prime mover. Referring to the figure 2.4,

$$T = P \tan \alpha$$

and
$$N = \frac{P}{\cos \alpha} \quad \text{...(2.7)}$$

It is seen that, the forces T and N depend upon the magnitude of eccentricity and angle of rotation θ, when P_d = constant. From the triangle OO_1A, using sine theorem,

$$\frac{R}{\sin \theta} = \frac{e}{\sin \alpha} \text{ or}$$

$$\alpha = \arcsin\left(\frac{e}{R} \sin \theta\right) \quad \text{...(2.8)}$$

When $e = 0$, $\alpha = \text{arc} \sin \theta = 0$ and when $e = e_{max}$, $\alpha = \alpha_{max}$. Also for the given eccentricity, the angle α depends upon angle θ. Combining equations 2.7 and 2.8,

$$T = P \tan\left[\arcsin\left(\frac{e}{R} \sin \theta\right)\right] \quad \text{...(2.9)}$$

From the above equation, it is evident, that T = 0 when θ = 0° or 180° and $T = T_{max}$ when θ = 90°. The resultant force arising due to all plungers T_{res} is due to the delivery pressure from individual plungers, ΣT. A graph $T_{res} = f(\theta)$ is drawn (figure 2.5) for 9 plungers. It is seen from the Figure that the combined force of all the plungers put together possess almost zero fluctuation, although individual plungers possess very high fluctuations.

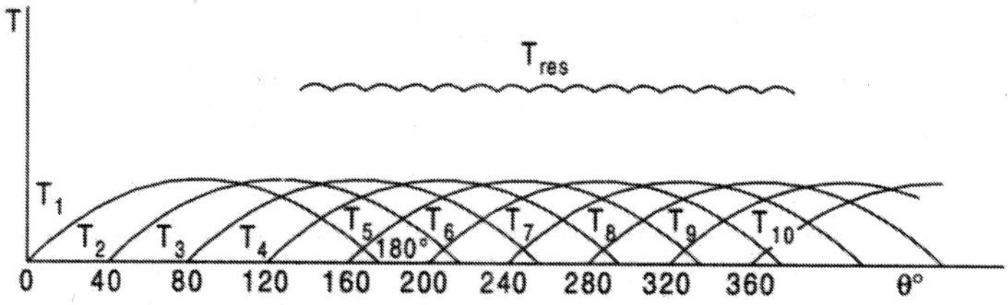

Figure 2.5. Change in the tangential force acting on the plunger and on the rotor

In the same manner, the relation between the force N with respect to θ and eccentricity can be established, by combining equations (2.8) and (2.9).

$$N = \frac{P}{\cos\left(\arcsin\left(\frac{e}{R}\sin\theta\right)\right)} \qquad ...(2.10)$$

From the equation, it is evident that N = P when 0 = 0° and 180°. The same thing can be proved when eccentricity is changed. Maximum value of N_{max}, is when θ = 90° as well as at maximum value of (*e*/R). The force $N_{res} = \Sigma N$, is independent of the angle θ, (refer the graph $N_{res} = f(\theta)$). Based on the above, the exact value of the force is calculated for the following components.

1. Base of the plunger sliding in the guide ring
2. Shaft and bearing
3. Guiding mechanism for the pump

2.4.1 Force Calculations for the Base or Top Rib of the Plunger

Maximum force K acting normal to the guide ring surface is determined from the instantaneous force N and inertia force on the plunger. From the equation 2.10, the force N is equal to

$$N = \frac{\pi d^2 p_d}{4\cos\left(\arcsin\left(\frac{e}{R}\sin\theta\right)\right)} \qquad ...(2.11)$$

in which, P_d is the major factor for the determination of the force N.

Referring to figure 2.2, in order to determine the inertia force, three types of accelerations, acting on the plunger namely relative acceleration j_1, lateral acceleration j_2 and rotating acceleration (coriolis component) j_3, are to be determined.

Relative acceleration j_1, can be calculated from the rate of change of relative velocity w_1, with respect to time *t*

$$j_1 = \frac{dw_1}{dt} = \frac{2\pi n.e}{60}\left[\left(1-\frac{e}{r}\cos\phi\right)\cos\phi + \frac{e}{r}\sin^2\phi\right]\frac{d\phi}{dt}$$

$$= \left(\frac{2\pi n}{60}\right)^2 e\left(\cos\phi - \frac{e}{r}\cos 2\phi\right) \qquad ...(2.12)$$

Lateral acceleration

$$j_2 = \omega^2(r+S) = \left(\frac{2\pi n}{60}\right)^2 (r+S)$$

$$j_2 = \left(\frac{2\pi n}{60}\right)^2 \left(\sqrt{r^2 + e^2\cos^2\phi + 2er} - e\cos\phi\right) \qquad ...(2.13)$$

In order to attain more reliable design, the centre of gravity of the plunger is assumed to be at point A (figure 2.2)

The coriolis component, $j_3 = 2\,\omega w_1$ ...(2.14)

This is directed perpendicular to the plunger axis but opposite to the direction of rotation of the rotor, if the plunger moves inward towards shaft axis and vice versa.

Total acceleration per unit mass $j = \bar{j}_1 + \bar{j}_2 + \bar{j}_3$, is the vectorial addition of these three individual accelerations. Total mass, m is equal to the mass of the plunger m_1, and mass of the fluid m_2 inside the cylinder. Therefore total force I is

$$I = (m_1 + m_2)j \qquad ...(2.15)$$

Due to variation of j_1, j_2, and j_3, the total acceleration j varies. Inertia force in an internal force of the system considered, hence work by this force is zero. Inertia force is considered, only for component design and for friction force calculations due to wear and tear of plunger and cylinder contact surface and not for power calculations. The coriolis acceleration of the inertia force is responsible for the wear of the sides of the cylinder and plunger parts due to compressive force acting on the sides. The total force I is resolved into two components, N and T (refer figure 2.2). These resolved components must be added vectorily with the resolved components of the forces due to fluid pressure N and T, and then by adding vectorily the total N and T, the total force I is determined. Since the theoretical calculation is a tedius and since such accuracy is not required, with considerable accuracy, inertia force can be calculated taking only lateral acceleration and neglecting relative and rotational accelerations. Taking $(r + S) = R$ and neglecting fluid mass inside cylinder. Total inertia force

$$I = m_1 R\omega^2$$

and

$$K = N + I \qquad ...(2.16)$$

2.4.2 Forces Acting on the Shaft and Bearings

Maximum pressure (P_{max}) occurs when more plungers are located in delivery area, instead of, in the suction area. Taking 5 plunger pump. Since the forces P_1, P_2, and P_3 (figure 2.6) act at a time, the total force P will be the vectorial summation of these three forces. The direction of P depends upon the angle θ (figure 2.6). If the delivery and suction are interchanged due to the change in the direction of rotation, corresponding direction of the resultant force also changes.

2.4.3 Force Calculation for the Pump Regulation

Flow regulation in these pumps are done by changing eccentricity '*e*', which is attained by shifting the guide ring along the horizontal axis *x-x* as well as keeping the guide ring at its position. The force S required is determined by (1) pressure underneath the plunger, (2) total inertia force of the plunger and (3) friction between plunger and cylinder, between guide ring and the base as well as between the horizontal plane of the guide ring and the corresponding plane in the casing or any operating guide.

Calculation of the total force S is difficult. Determination by graphical method is quick, simple and sufficiently accurate. The effect due to centrifugal force and friction is neglected. Under this condition the total force S is directly proportional to the number of

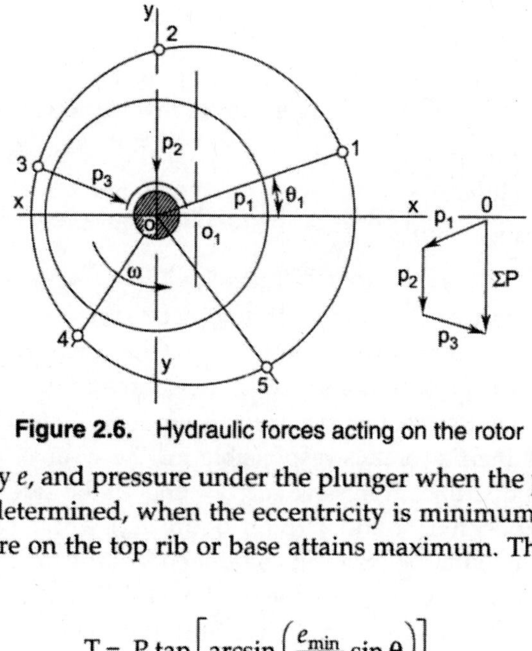

Figure 2.6. Hydraulic forces acting on the rotor

cylinders, eccentricity *e*, and pressure under the plunger when the plunger is at delivery stroke. The force is determined, when the eccentricity is minimum *i.e.*, $e_{min} = me$. At this condition the pressure on the top rib or base attains maximum. The condition the force T is calculated as

$$T = P \tan\left[\arcsin\left(\frac{e_{min}}{R}\sin\theta\right)\right] \qquad ...(2.17)$$

From the resultant values of $P_{res} = \sum P$ and $T_{res} = \sum T$, the resultant force $\sum N$ is obtained, which acts normal and directed towards the outer side of the guide ring, (figure 2.7). The net force N is represented by vector OA, which is resultant from 5 plungers. As mentioned earlier $\sum T$ is independent of θ, maximum value of $\sum N$, projected on the axis *x-x*, can be determined, which is obtained by rotating the vector OA by an angle $(\lambda - \theta_1)$ (θ_1 belongs to the condition of plunger 3 at maximum delivery pressure).

Likewise, by geometrically constructing, maximum force $\sum N_x = S$ is determined, which moves the guide ring and the horizontal force $\sum N_y$, $\sum N_x$, and $\sum N_y$ are the two forces necessary to determine the dimension of guiding mechanism.

2.5 LOSSES, EFFICIENCY, PERFORMANCE, CHARACTERISTICS OF THE PUMP

The losses and efficiencies are calculated in the same way as for the reciprocating pump. There is no air vessal. Liquid flows through the plunger and cylinder by travelling a minimum distance. Hence hydraulic losses are negligible. Volumetric and mechanical efficiencies are taken from experimental results.

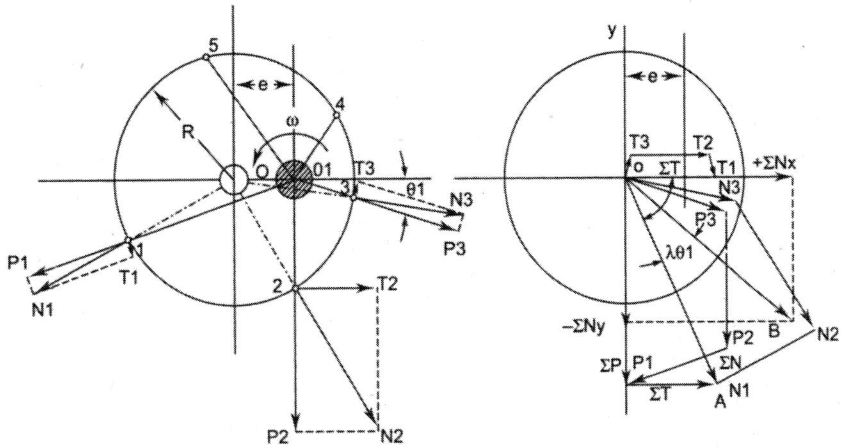

Figure 2.7. Vector diagram to calculate the forces for guiding mechanism

2.5.1 Volumetric Efficiency

Volumetric efficiency η_v varies from 50% to 98%. The losses are due to leakage through clearance and due to incomplete filling of liquid inside the cylinder space. Volumetric efficiency decreases, with the increase in pressure inside the cylinder, the amount of clearance between plunger and cylinder, speed and viscosity of the liquid. It also decreases when eccentricity decreases due to the reduction of flow rate and constant leakage flow. In figure 2.8, the change of volumetric efficiency with respect to the delivery pressure is given./2/

2.5.2 Mechanical Efficiency

Mechanical efficiency lies between 75% and 97%. It is due to friction between plunger and cylinder, due to friction between top rib and guide ring , disc friction losses *i.e.*, friction losses due to rotor immersed inside liquid and losses in bearing and other parts. It varies depending upon the construction. It depends upon the delivery pressure also. When the pressure is below 65 to 75 ata, the loss due to friction considerably increases and hence efficiency decreases. Variation of η_m as a function of delivery pressure is given in figure 2.8.

2.5.3 Overall Efficiency

Overall efficiency, $\eta = \eta_v \cdot \eta_m \approx 0.4$ to 0.5. It purely depends upon the quality of manufacture, losses and the operating point. In figure 2.9, the variation of overall efficiency as a function delivery pressure is illustrated./2,22,23/

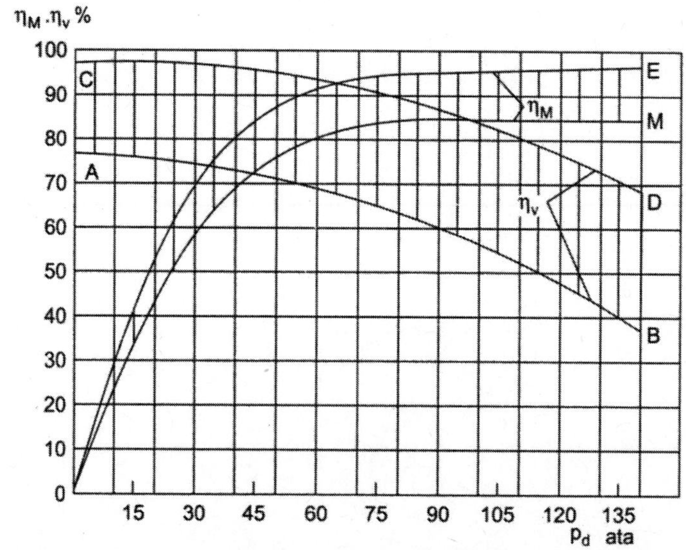

Figure 2.8. Variation of volumetric and mechanical efficiency with respect to delivery pressure

2.6 DETERMINATION OF BASIC DIMENSIONS

The basic dimensions for a radial piston pump are

1. Plunger diameter 'd' and
2. Stroke length 'S'

All other dimensions are determined based on these two dimensions. The strength or the determination of thickness is determined based on the delivery pressure and inertia force.

For the given value of flow rate (Q_a) m^3/hr or lit/min and delivery pressure 'p_d' ata or kg/cm^2 or the total delivery head H in meters and the property of the liquid such as temperature, viscosity, density and vapour pressure, the basic dimensions are determined.

2.6.1 Selection of Speed (n)

Increase in speed, reduce the size of the pump, but leads to difficulty in manufacture and consequent dependability in operation. The volumetric efficiency decreases at a higher rate especially at very high speeds. The plunger movement is so fast, the liquid does not occupy complete space available in the cylinder, as a result, the liquid from delivery enters the space. Flow rate considerably reduces. Normal recommended speed n = 550 to 1200 rpm.

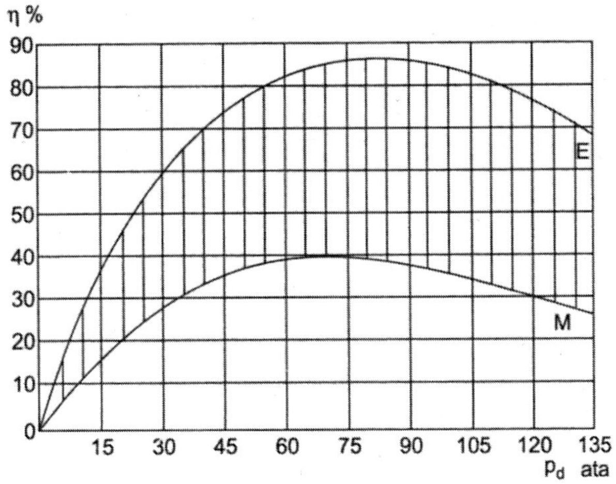

Figure 2.9. Variation of overall efficiency with respect to delivery pressure

2.6.2 Number of Cylinders (Z)

The number of cylinders Z vary from 7 to 9, in one rotor. For higher flow rate instead of increasing the stroke length or increasing the number of cylinders, another parallel pump is attached to the same shaft.

2.6.3 Selection of S/d Ratio

The stroke diameter ratio S/d = m is selected between 0.8 and 1.13. It is to mention that S = 2e.

2.6.4 Selection of Plunger Diameter d

If *i* is the number of pumps, connected to a single shaft and Z is the number of cylinders in each pump,

$$Q = Q_a = \eta_v \frac{\pi d^2}{4} .i.m.nZ$$

or

$$d = \sqrt{\frac{4Q}{\eta_v\ \pi\ .i.m.nZ}} \quad \text{...(2.18)}$$

Due to the presence of coriolis force due to uneven distribution of forces along the plane of rotation and friction in the guide ring, there will be an erosion in the cylinder surface. In order to obtain uniform wear at the cylinder surface as well as to reduce the pressure intensity on the cylinder surface, the contact length a in figure 2.10, between the cylinder and plunger, when the piston is at a maximum height or at the end of full stroke length will be

$$a \approx (1.8 \text{ to } 2.0)\, S \quad \text{...(2.19)}$$

Excess length increases friction and reduced length creates uneven movement.

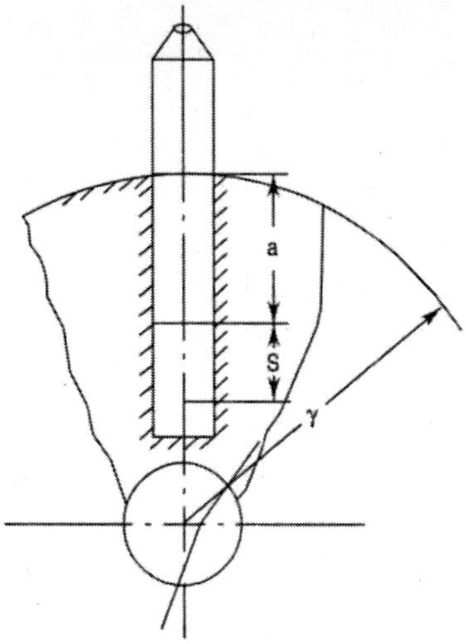

Figure 2.10. Contact surface for radial piston pumps

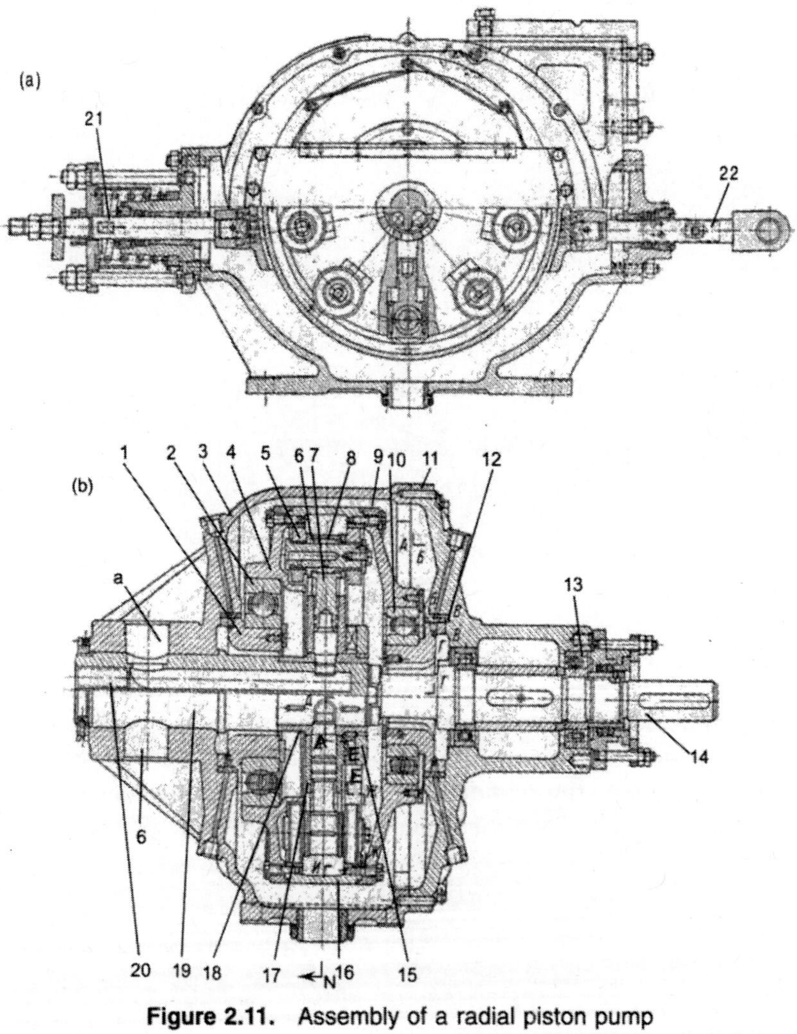

Figure 2.11. Assembly of a radial piston pump

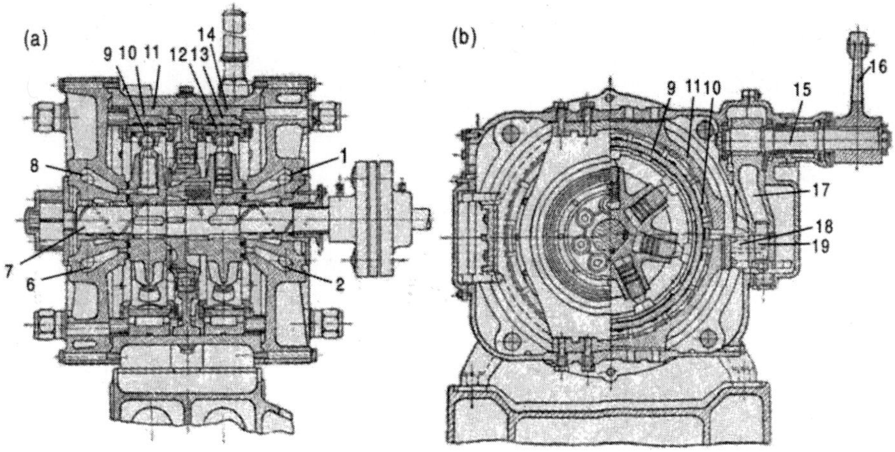

Figure 2.12. Front and side view of a radial piston pump

CHAPTER 3

AXIAL PISTON PUMPS

3.1 PRINCIPLE AND CONSTRUCTION

In axial piston pumps, the plunger is located either parallel or at an angle to the axis of the swash plate. Axial piston pumps weigh less and possess smaller size, when compared to the radial piston pumps, for the same flow rate and delivery pressure. But Volumetric efficiency is less in axial piston pumps, when compared with radial piston pumps.

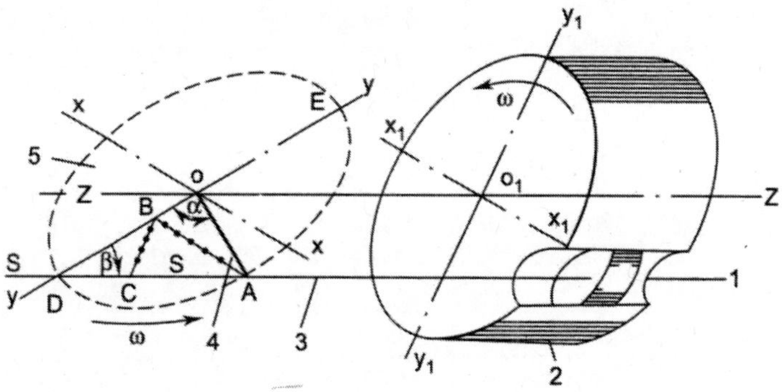

Figure 3.1. Kinematic Diagram for an axial piston pump

In order to understand the principle of action of the axial piston pump, a kinematic diagram is illustrated in Fig. 3.1, Rotor 2 has five to nine circular holes, called slots, symmetrically located around the circumference. Equal number of plungers, 1, one in each slot, reciprocates. The connecting rod 3 is connected to a swash plate 5, and to the plunger with a spherical ball type joint at both ends, on the plane *xoy*. The line OA in the plane *xoy* represents a curved contact path 4. When rotor rotates parallel to horizontal axis passing through O_1 and the swash plate 4 with respect to the point O, on the plane *xoy* along the direction of rotation as indicated in the Fig. 3.1, each plunger 1 undergoes one complete rotary and one complete reciprocating motion by which one complete suction and delivery strokes are carried out. Reciprocating motion for each revolution is attained by the inclination of the swash plate with respect to rotor which is vertical. Flexible movement is attained by the ball and socket groove arrangement. The plane *xoy* has to be considered as the spread over surface kept perpendicular to the surface $x_1o_1y_1$ in such a way that the plunger and the connecting rod 3 are in the same plane of *xoy*. If the curved groove 4 rotates with respect to the point O keeping the rod 3 always parallel to the axis Z, along with the rotor 2, the plunger makes one complete reciprocating motion. Taking $\angle AoB = \alpha$ and radius of groove OA = R

$$BD = y = OD - OB = R - R\cos\alpha = R(1 - \cos\alpha) \quad ...(3.1)$$

If the plane surface xoy is considered as a three dimensional surface, the circle 5 must be inclined at some angle $\angle BDC = \beta$ with respect to horizontal, *i.e.*, the swash plate 5 is inclined at an angle β to the horizontal axis but it rotates with respect to the horizontal axis OO_1.

The Kinematic principle of Axial Piston Pump (also called as piston type rotary pump of axial type) is similar to that of regular piston pump *i.e.*, conversion of rotary motion into reciprocating motion by Simple Harmonic Motion. In Axial piston pump, along with the cylinder shifts vertically parallel to itself, and at the same time, the piston reciprocates relative to the cylinder due to the provision of inclined swash plate (Fig. 3.2).

During the rotation of an angle ϕ by the inclined swash plate, the piston moves vertically along with the cylinder by an angle ϕ and at the same time axially moves away from the dead centre by a distance *x* which is equal to

$$X = R(1 - \cos\phi)$$

where, R is the radius of swash plate.

Let the inclined swash plate be rotated in the inclined position about the axis *y-y* through angle β (Fig. 3.3) and the cylinder with piston is shifted parallel to its own axis of rotation, along with the plane of rotation of the shaft of the swash plate. Point A is also displacing along the axis *y-y*. The end of the piston, where it touches the inclined swash plate in the form of either spherical ball or fitted with a roller bearing, lies on the circumferential groove the cross section of which is in the form of arc of the circle on the swash plate, on which the spherical end of piston is positioned and moving. Assuming with a point positive contact at all times of movement between spherical ball of the piston and the groove in the form of arc of the circle on the swash plate, during the rotation of swash plate through an angle β, the cylinder must shift to the distance which is the projected distance of the arc of the circle, developed by the point A resolved to the plane perpendicular to the cylinder axis *y-y*. During this rotation, the piston shifts to an amount cos β. When β = 0, the flow will be plane flow. Under this condition, the shifting of the piston is as per the equation (Fig. 3.2).

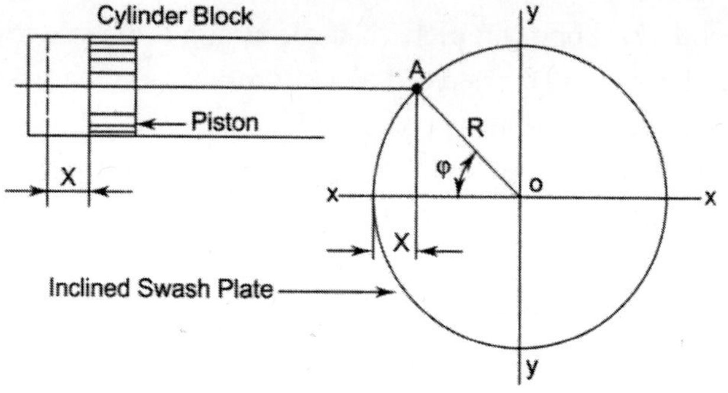

Figure 3.2. Scheme of working of Piston in Axial Piston Pump

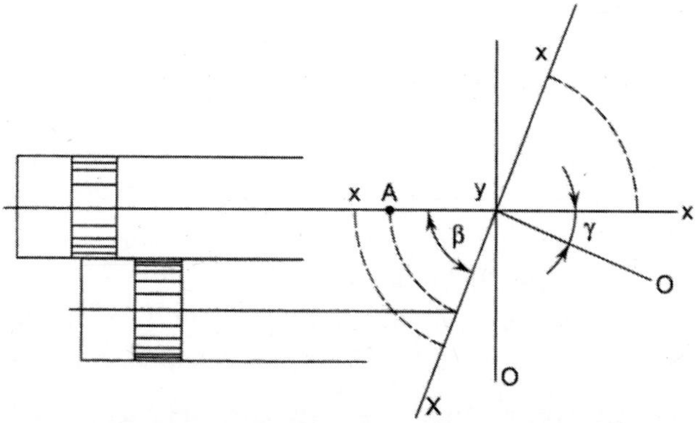

Figure 3.3. The analysis of principle of working of piston of Axial Piston Pump

$$x = e\,(1 - \cos \phi)$$

When the swash plate shaft plane is shifted such that $\beta = 90°$, the shifting of the piston with respect to the cylinder will be equal to Zero, since $\cos 90° = 0$. Hence it can be said that, when the swash plate rotates by an angle β, the shifting of piston with respect to the cylinder can be written as

$$X = R\,(1 - \cos \phi) \cos \beta \qquad \text{...(3.2)}$$

If a number of cylinders are located around the circumference of the cylinder block, but the swash plate is moved with the disc, the axis of which is tilted related to the cylinder axis by an angle γ in such a way that $\beta + \gamma = 90°$, and the plane of rotations of the disc coincides with the plane of rotation of the swash plate, we get axial piston pump, reciprocating axially with respect to the cylinder block and the piston simultaneously rotating with cylinder block (Fig. 3.4).

An axial piston pump consists of stationary distribution disc 1, rotating cylindrical block or rotor 2, piston 3, connecting rod 4 and inclined disc or plate 5, called as swash plate. The contact of 4 with 5 is a spherical ball positioned on a circular arc of a circular groove arranged on a swash plate of the circular form.

Window openings 6 in the form of arc of a circle (Fig. 3.2) are provided on the distribution disc, through which the fluid enters to and leaves from the spaces between the piston and cylinder due to the rotation of disc inclined 5 with the center of shaft 7, but disc rotates with cylinder block 2 with respect to the axis O-O.

Practically, the angle of rotation of the inclined disc γ is 12° to 15° and goes upto 30°. In order to reduce the flow fluctuations, total number of cylinders must be infinity. But practically number of cylinders is taken from 5 to 9.

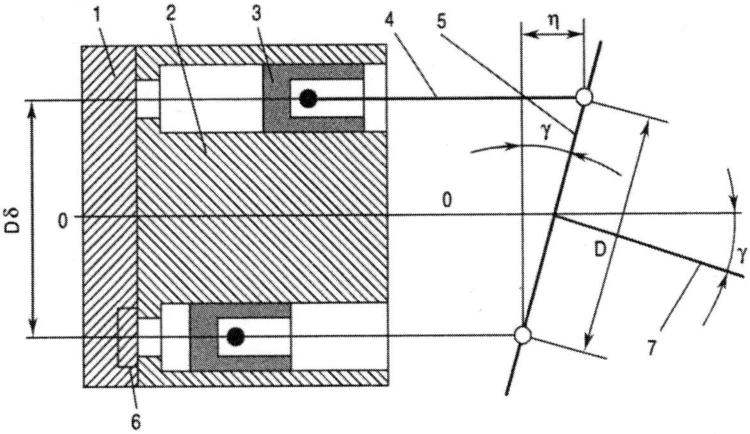

Figure 3.4. Scheme of Axial Piston Pump

3.2 KINEMATICS OF THE PUMP

If the inclined disc 5 (Fig. 3.4) rotates about its own axis at constant angular velocity, and the cylinder block along with the piston is rotating about the axis O-O the piston makes a reciprocating motion. It returns back to its original position after making one full stroke relative to the cylinder (Fig. 3.3). It can be seen that the rotation of point A is obtained by the projection of the arc of the circle on the plane perpendicular to the cylinder block axis. The reciprocating motion of the piston is described by the motion of inclined plate the circle with the axis O-O. That's how the connecting rod between piston and the inclined swash plate is getting an oscillating or reciprocating motion relative to the cylinder axis. Since the angle of tilt or movement of the disc and its rotation is very small, the oscillation of the connecting rod can be neglected and it can be assumed that they move, during the rotation of the inclined disc, parallel with respect to axis. Its axis coincides with the cylinder axis.

During the rotation of the inclined disc by an angle ϕ the piston displace S to a distance of 'x'. The relative velocity of the piston will be equal to the displacement x during the time 't'

$$V_p = \frac{dx}{dt} = R\cos\beta\sin\phi\,\frac{d\phi}{dt}$$

or

$$V_0 = \omega R\cos\beta\sin\phi \qquad ...(3.3)$$

When

$$\beta = 90 - \gamma,\ \cos\beta = \sin\gamma$$

$$v_p = \omega R\sin\gamma\sin\phi \qquad ...(3.4)$$

From the above equation, it is evident that the piston moves with sinusoidal law. When $\phi = 0$ the relative velocity $v_p = 0$. Maximum relative velocity of the piston v_p occurs when $\phi = 90°$.

$$V_{pmax} = \omega R \sin \gamma$$

Since the motion of the piston is a reciprocating motion, the acceleration of the piston.

$$a_p = \frac{dv_p}{dt} = \omega^2 R \sin \gamma \cos \phi \qquad ...(3.5)$$

Correspondingly, acceleration of the piston in relative motion, changes, as per cosine law. When $\phi = 0$ the acceleration is relative motion is maximum

$$a_{pmax} = \omega^2 R \sin \gamma \qquad ...(3.6)$$

When $\phi = 90°$ acceleration is minimum. At this time there is no acceleration $a_p = 0$. When $\phi = 180°$ the piston acceleration is

$$a_0 = \omega^2 - R \sin \gamma \qquad ...(3.7)$$

The acceleration of the piston will be positive when (swash plate) the inclined plate in inclination is between 0 – 90°. When ϕ changes from 90° to 180° and between 180° and 270° the acceleration is negative. When the inclination is from 270° to 360° the acceleration is again positive.

This shows that the inertia force in relative motion changes from positive to negative in one revolution. During Ist quarter (0° to 90°) and the IVth quarter (270° to 360°) of the relative motion of the piston, the swash plate moves in positive direction and during IInd and IIIrd quarter (90° to 270°) of the relative motion, the inertia force is in the opposite directions (negative).

At the same time, the piston also moves at constant angular velocity. This rotary movement and corresponding acceleration of the piston with respect to the cylinder axis is equal to the centripetal acceleration (radially inverse)

$$a_p = \omega^2 R_c \qquad ...(3.8)$$

where R_c is the radius of the cylinder block.

Inertia force, in this motion, is directed towards the centre of the cylindrical block in radial direction (perpendicular to the axis of the cylinder).

This inertia force increases the friction between piston and cylinder. The corioliss force is equal to zero in this condition since angle, subtended between cylinder block centre and the direction of the relative velocity of the piston movement, is zero.

3.3 CALCULATION OF FLOW RATE (Q)

The flow rate is determined by the movement of the swash plate 5 in axial direction, when it is rotating with OO_1, as axis of rotation. Referring to Fig. 3.1

$$DC = h = BD \cos \beta = R(1 - \cos \alpha) \cos \beta \qquad ...(3.9)$$

where h is the stroke length

The velocity $$v = \frac{dh}{dt} R \cos \beta \sin \alpha \frac{d\alpha}{dt}$$

Since $$\frac{d\alpha}{dt} = \omega, v = \omega R \cos \beta \sin \alpha \qquad ...(3.10)$$

Flow in each cylinder $q_{th} = v.A = \omega R \cos\beta \sin\alpha \dfrac{\pi d^2}{4}$ ...(3.11)

Where A is the area and d is the diameter of the plunger. Equation (3.4) indicates, that the flow rate is in sinusoidal form. The stroke length $h = 2\,R \cos\beta$. The quantity of flow, from one plunger for one revolution $q_{th} = h.A$. If there are z number of plungers, and n is the speed in rpm the total flow rate, Q_{th}, is

$$Q_{th} = Z.n.q_{th} = \frac{\pi d^2}{4}\ \omega R \cos\beta \sin\alpha.Z.n \qquad \text{...(3.12)}$$

The flow rate can be changed by changing the stroke length h, which is done by changing the inclination of swash plate β. The speed of these pumps varies from 1500 to 5000 rpm.

During one revolution of the inclined disc, the piston completes one full up and down stroke. The flow is equal to

For one revolution $q = \dfrac{\pi d^2}{4} .h.$

Where, h is the stroke of the piston and d is the diameter of the piston. If there are z number of pistons and speed is n rpm the flow rate will be

For one revolution $q = \dfrac{\pi d^2}{4} .h.z.n\ m^3/\text{min}$

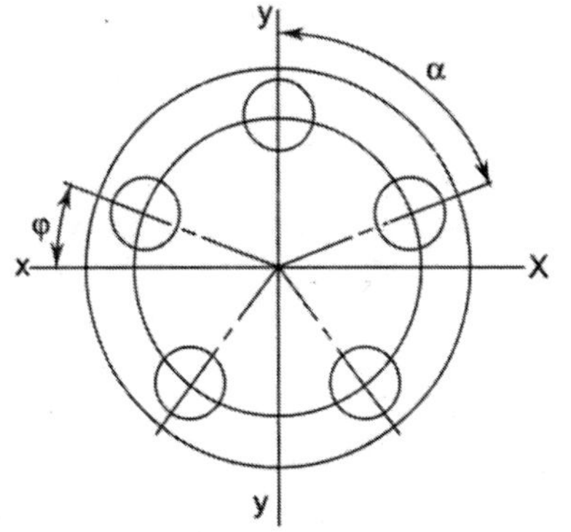

Figure 3.5. Position of cylinders in the block

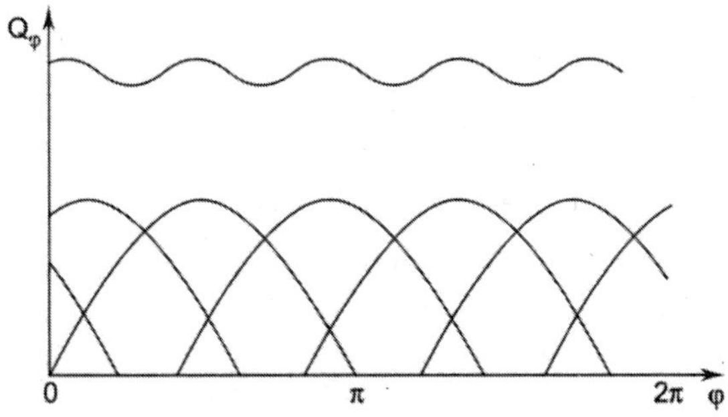

Figure 3.6. Graph explaining some total of flow in pumps

Referring to Fig. 3.5

To show in figure Stroke $h = D \sin \gamma = D_c \tan \gamma$.

When D_c is the diameter of the cylinder block

D is the diameter of the inclined disc

So,

$$Q = \frac{\pi d^2}{4}.ZnD \sin \gamma$$

$$= \frac{\pi d^2}{4}.ZnD_c \tan \gamma \quad \text{...(3.13)}$$

Instantaneous flow from one piston, using (3.4) is

$$q_\phi = \frac{\pi d^2}{4} \omega R \sin \gamma \sin \phi$$

Total flow of the pump

$$Q_\phi = \frac{\pi d^2}{4} \omega R \sin \gamma [\sin \phi + \sin(\phi + \alpha) + \ldots + \sin(\phi + m\alpha)] \quad \ldots (3.14)$$

Where $\propto$ is the angle between two successive cylinders

$$\propto = \frac{2\pi}{z} \text{ (Fig. 3.5)}$$

Using equation 3.14, a graph is drawn Fig. 3.5 shows a graph between the quantity Q_ϕ and the displacement angle ϕ for each pump and also for the sum total quantity coming out from all z number of pistons of the pump when running with respect to the inclination of the inclined disc ϕ.

Equation 3.14 is same as the equation for radial piston pumps. All other aspects such as pulsating flow, the coefficient of unequalness of flow are same for axial and radial piston pumps.

3.4 FORCES ACTING ON PUMPS

The pressure acting on the piston reduces due to friction, which reduces the power or moment available from the shaft to the piston.

If the diameter of the piston is d and the internal pressure in piston is p, the force F is equal to

$$F = p.\frac{\pi d^2}{4}$$

If the connection between the connecting rod end and the circular groove in the inclined disc is a free point or line contact ball and socket type, or roller bearing fitted contact type the resolved component of the force F in normal N and along the direction are, as shown in (Fig. 3.7). Values of N and P are

$$N = F \cos \gamma$$

and

$$P = F \sin \gamma$$

The pressure P again can be resolved in two directions *i.e.*, parallel to axis and perpendicular to axis of the inclined disc (located at an angle ϕ).

The moment M = TR, where R is the disc radius and T is the force perpendicular to the axis of rotation of inclined plate since $T = P \sin \phi$, the movement to rotate the inclined plate

$$M = F\,R \sin\gamma \sin\phi \qquad ...(3.15)$$

or

$$M = F\,R_c \tan\gamma \sin\phi \qquad ...(3.16)$$

where R is the radius of inclined plate

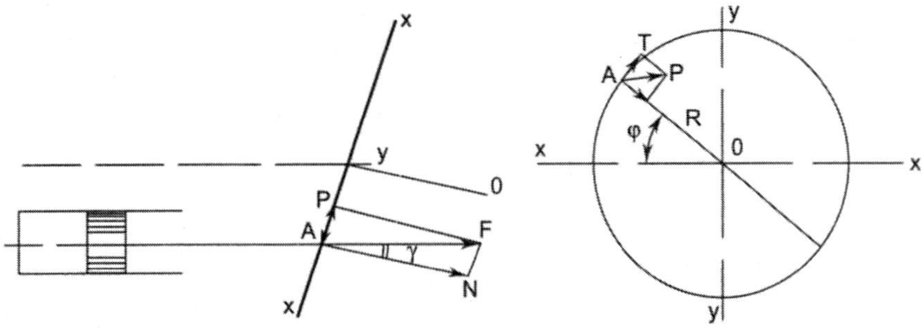

Figure 3.7. Forces acting on Axial Piston Pump

The total torque at the shaft is the sum of the moments given by the pistons, located at the delivery area.

$$M_c = \Sigma M = FR \sin\gamma\,[\sin\phi + \sin(\phi + \alpha) + \ldots + \sin(\phi + m\alpha) \qquad ...(3.17)$$

This torque also changes as that of the flow, which is evident, by comparing equations (3.14 and 3.17). When $Q = Q_{max}$, the torque is maximum and vice versa. In other words, when the swash plate is inclined and is equal the pulsation angle, the total torque is maximum. If the suction side is also under pressure inspite of positive value at the delivery side, the resultant moment calculated including the opposite pressure on the suction side, is

$$M_R = M_c - M_n$$

Where M_n is the torque given by the opposite pressure from the suction side. In Fig. 3.8 illustrated, the axial piston pump with free compression with the inclined swash plate by the piston. It should be noted that, when the piston is connected to the swash plate with the ball and socket arrangement it is a free, connecting when in contact with the inclined plate

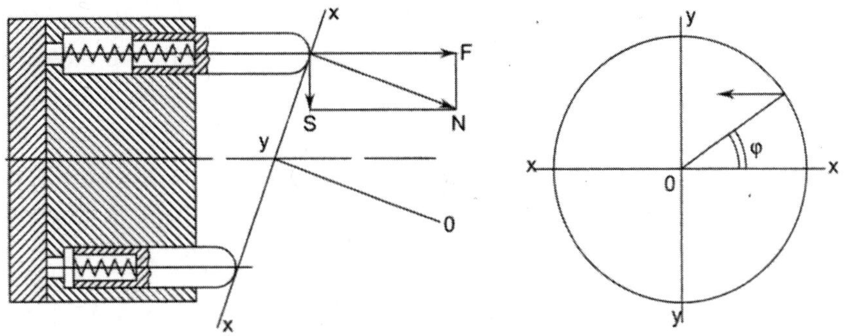

Figure 3.8. Scheme of Axial Piston Pump with piston when free and under compression of the disc

Actually the force N given by the piston on the inclined plate is resolved into two components. One, the force F directed along the piston and another, S directed perpendicular to the piston axis.

The force S gives a moment relative to the axis of the cylindrical block. Under this condition (Fig. 3.8)

$$M = S R_c \sin \phi$$

But $$S = F \tan \gamma$$

Therefore $$M = F R_c \tan \gamma \sin \phi \quad ...(3.18)$$

Neglecting the shift of location on the force N relative to the piston axis, the turning moment on the inclined plate is determined as follows. The point, tangent to the piston and the inclined plate coincides with the piston axis. So the force N gives, relative the axis *y-y*, a moment, which tries to change the angle of rotation of the plane of the rotating disc, and relative to the axis *x-x*, a moment giving a force or load on the bearings located on the axis of rotation of the rotating disc.

From Fig. 3.8 it is evident that relative to axis *y-y*, the turning moment given by the force N will be equal to the resolved component of the force N on the area, which is equal to R cos ϕ

$$M_y = N R \cos \phi$$

Since, $$N = \frac{F}{\cos \gamma}$$

$$M_y = \frac{F}{\cos \gamma} R \cos \phi$$

Since, $$R = \frac{D_c}{2 \cos \gamma}$$

$$M_y = \frac{F}{\cos^2 \gamma} \cdot \frac{D_c \cos \phi}{2} \quad ...(3.19)$$

From equation 3.19 it is seen than when $\phi < 90°$, the moment M_y is clockwise and when $\phi > 90°$ it is in anticlockwise. Maximum value of the rotating moment is when $\phi = 0$. When $\phi = 90°$ this moment is zero. When $\phi = 180°$, the value of this moment is again maximum but directed towards the opposite direction.

However if at the delivery side, the number of cylinders are even and the piston are symmetrically located and rotated with respect to axis *y-y* the sum of the turning moment is zero.

If number of cylinders or piston at the delivery side is odd $(m + 1)$ they will be located unsymmetrical to the axis *y-y*. The resultant or the total turning moment on the pump disc is equal to

$$\Sigma M_y = \frac{F}{\cos^2 \gamma} \cdot \frac{D_c}{2} [\cos \phi + \cos (\phi + \alpha) + \ldots + \cos (\phi + m\alpha)] \quad ...(3.20)$$

The total turning moment will be zero if

$\cos\phi + \cos(\phi + \alpha) + \ldots + \cos(\phi + m\alpha) = 0$

From which $$\tan\phi = \frac{1 + \cos\alpha + \ldots + \cos(m\alpha)}{\sin\alpha + \ldots + \sin(m\alpha)}$$

Naturally the sum total of turning moment is zero, when the angle of rotation of the inclined disc is equal to the angle of pulsation which occurs, when the quantity or flow rate is maximum.

The inertia force of the piston reacts on the sum total of the resulting moment. Under non moving condition of the inclined disc but rotating cylindrical block, the inertia force relative to the moment plays as major role, which can be determined from equation (3.11).

This force during I[st] half of the delivery and during IInd half of suction is positive (+ve) and directed on the inclined disc and tries to turn relative to axis *y-y*. During IInd half of the delivery and I[st] half of suction this inertia force in relative direction is negative (–ve) and is directed opposite and tries to remove piston from the swash plate.

That's why the inertia force, during the position piston of IInd half of delivery and I[st] half of suction will not create any turning moment relative to the axis *y-y*.

If the mass of the piston is *m*, the moment given by the inertia force relative to the axis *y-y*.

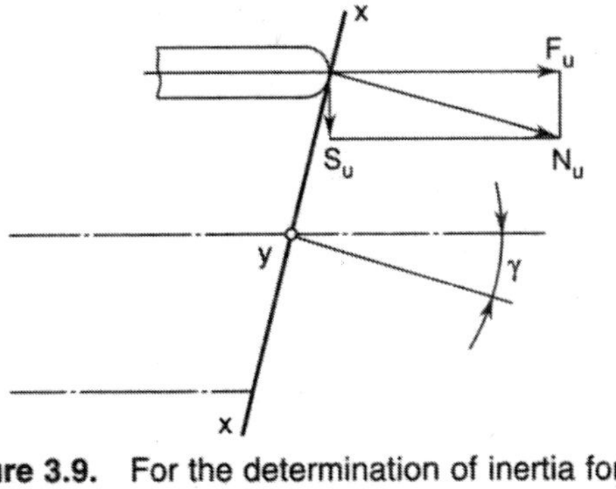

Figure 3.9. For the determination of inertia force

$$M_{yu} = \frac{F_u}{\cos\gamma} R\cos\phi$$

where $$F_u = m\omega R \sin\gamma \cos\phi$$

therefore $$M_{yu} = m\omega^2 R^2 \tan\gamma \cos^2\phi$$

$$= \frac{m\,\omega^2 R_c^2 \tan\gamma}{\cos^2\gamma} \cos^2\phi \qquad \text{...(3.21)}$$

If during the I[st] half of the delivery are located number of pistons *b* and during IInd half of the suction are located number piston *c*, then the sum total of the resultant moment of the inertia form relative to the axis *y-y* will be

$$\Sigma M_{yu} = m\omega^2 R^2 \tan\gamma \{\cos^2\phi + \cos^2(\phi + \alpha) + \ldots + \cos^2[\phi + (\beta - 1)\alpha] + \cos^2(\phi + (z - c)\alpha] + \ldots + \cos^2[\phi + (z - 1)\alpha]\} \qquad \text{...(3.22)}$$

Actually the effect of the shift of the point of contact belong to the force N and N_u will also react on the turning moment relative to the axis *y-y*. This shifting of force N and N_u takes place on the spherical ball surface of the piston, which can be determined from the geometric condition. From Fig. 3.10 it is evident that, the shifting of the point of contact of the force N goes on changing the actual turning moment. If r_0 is the piston radius and the value of the shift is Δ, then

$$\Delta = r_0 \tan \gamma$$

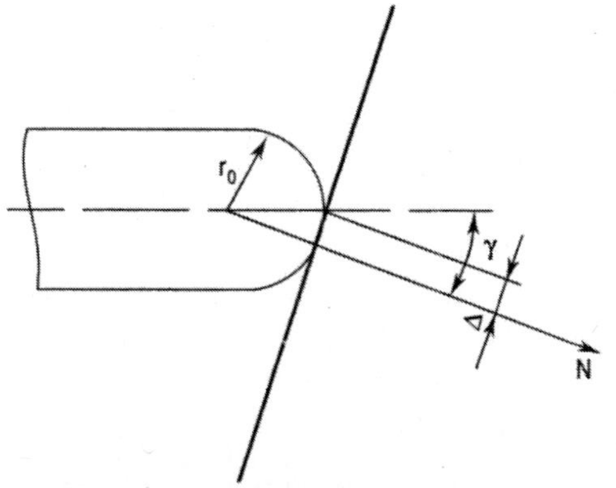

Figure 3.10. Shifting of the point regarding the force N

Referring to Fig. 3.8 it is easy to find that for the piston located above the axis *y-y* the actual turning moment will be less by an amount Δ and viceversa. So, for the piston located above the axis *y-y* the resultant will be (R cos ϕ – Δ) and for the piston located below the axis *y-y* it will be (R cos ϕ + Δ).

Relative to the axis *x-x* the moment of this force will act on the bearing of the inclined disc. From Fig. 3.8 we can written as

$$M_x = \frac{F}{\cos \gamma} R \sin \phi$$

If there are $(m + 1)$ pistons located in the delivery side, then the total moment relative to axis *x-x* is determined from the equation

$$\Sigma M_\alpha = \frac{F}{\cos \gamma} R [\sin \phi + \sin (\phi + \alpha) + \ldots + \sin (\phi + m\alpha) \quad \text{...(3.23)}$$

All the above forces are determined, when the inclined disc is not rotating. If the inclined disc is rotating, the sum total of the rotating moment influences on the centrifugal force of the inertia force of the rotating disc, which gives a moment relative to the axis *y-y*, which tries to tilt the inclined disc in position, where the angle β is equal to 90°. If the disc weight is G and the distance from the axis *y-y*, to the centre of gravity of the inclined disc of the half disc is 'l' than the moment given by the centrifugal inertia force relative to axis *y-y* (Fig. 3.11) is equal to

$$M_{centrifugal} = \frac{G}{g} bl\omega^2 \quad \text{...(3.24)}$$

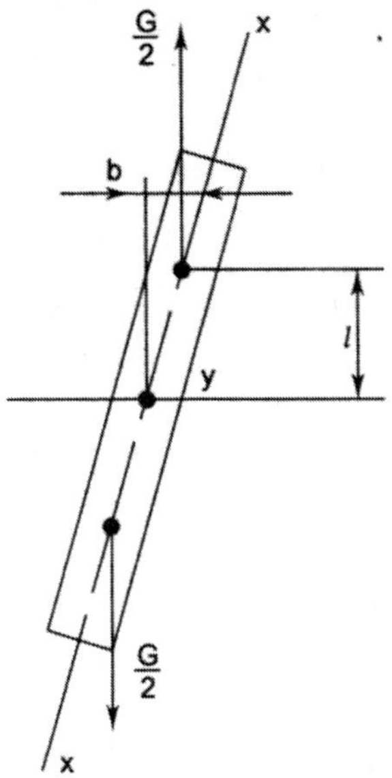

Figure 3.11. For the determination of moment due to centrifugal force

3.5 THE DISTRIBUTION DISC

Entry and Exit of liquid in axial piston pumps, are considered relative to the condition that, the distributed disc is either nonmoving or moving.

The first is adopted in pumps where the cylindrical block is rotating and the IInd in pumps, where cylindrical block is stationary but inclined plate is rotating. In Fig. 3.12 is illustrated a non-moving distribution disc. With the help of the canal and the window made in disc, the suction and delivery pipes are connected with the pump cylinder. Between canals a width '*b*' is provided to identify the suction and delivery. During the rotation of the cylindrical block, the cylindrical holes are connected to either suction or to the delivery. The suction and delivery ports are determined based on the direction of rotation of the cylindrical block. When the rotating direction of the cylindrical block is reversed, the suction and delivery parts are inter charged.

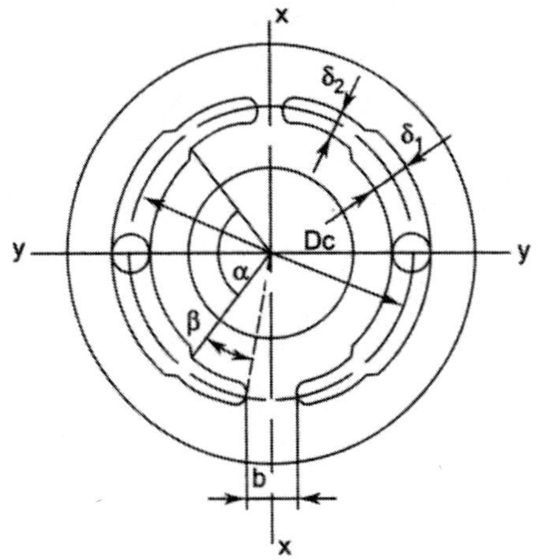

Figure 3.12. Nonmoving distribution disc

In order to reduce leakage, the side surface of the cylindrical block is firmly fixed with the distribution disc. Both are firmly in contact at all times.

Under ideal condition of contact between the distributed disc and the cylindrical block *i.e.*, there is no leakage, A force 'F' from the distribution disc towards the cylindrical block will be acting which is

$$F_1 = p\,(\pi D_X \frac{\alpha}{360}\,\delta_1 + 2\pi D_X \frac{\beta}{360^\circ}\,\delta_2)$$

where p is the delivery pressure

This force F, tries to remove the distribution disc away from the cylindrical block, but the delivery pressure compresses towards the disc. Since number of pistons in axial piston pump are odd in number the area of delivery pressure may lie either $\frac{Z+1}{2}$ or $\frac{Z-1}{2}$ pistons. So the minimum force the block acts on the disc is

$$F_2 = p\,\frac{\pi d^2}{4}\,\frac{(Z-1)}{2} \qquad \text{...(3.25)}$$

Actually due to the clearance between the distribution disc and the cylindrical block, the pressure p changes in the canals of the disc from p in the canal to zero at the periphery side of the disc. This force F_{1d} tries to remove the block from the disc since the force F_{1d} is more than F_1 due to leakage in the clearance.

Assuming linear change of pressure variation from the canal, to the outer periphery in the clearance between the disc and the block.

$$F_{1d} = F_1 + 0.5\,p\,(A_1 + A_2) \qquad \text{...(3.26)}$$

where A_1 and A_2 are the inner and outer area of contact of the disc and block.

In order to have a consistency in pressure, the pressure F_2 must be greater than F_{1d}. For normal clearance between disc and block.

$$F_2 = 1.1\,F_{1d}$$

In order to reduce the force F_{1d}, the contact surface of the disc is made with the pressure relieving grooves. By this the fluid leaves through the groove and goes through to the casing. Due to this the relief between block and disc on the delivery canal and at half ring, the pressure is reduced at the delivery canal.

By making the groove on both sides of the disc, the outer ring *i.e.*, the contact surface between the canal and outer side of the pressure relieving canal, is made a little wider in order to reduce the effect of centrifugal force due to piston inertia on the leakage flow.

If δ_1 is the width of the canal on the disc and is made equal to the diameter d of the gap in the cylinder of the pump and the width δ_2 is equal to δ_1 then the angle $\beta = ½\,\alpha$, the width of the space between the area suction and delivery grooves $b = d_0 + 0.6$ mm, where d_0 is the diameter of the hole in the block connecting the cylinder hole with the surface of the disc (Fig. 3.13).

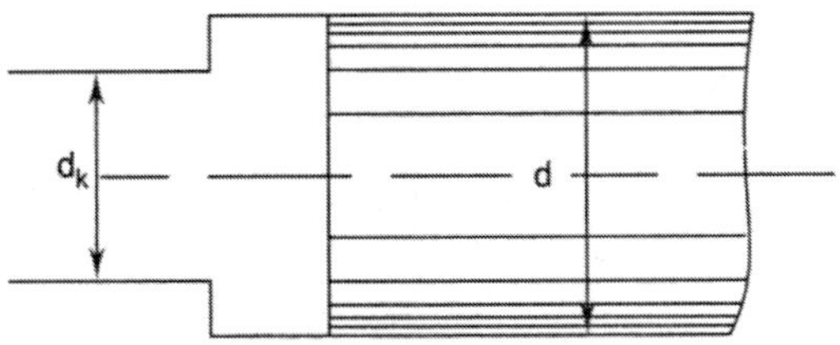

Figure 3.13. Scheme indicating the relation between the diameter of the hole in block and in the cylinder

Since the pressure distribution in the side ways of the fluid with the rotating block is in the form of radially distributed pressure and is same as that in radial piston pumps.

The radial pressure distribution is technically simple that could not be relieved with respect to the leakage. It can be assumed that for a good pressure distribution at the side, a high volumetric efficiency is allowed.

For a free approach and exit of liquid the side distribution disc is attached, with the nonmoving block and rotating swash plate, which rotates together with inclined swash plate at the same speed (Fig. 3.14).

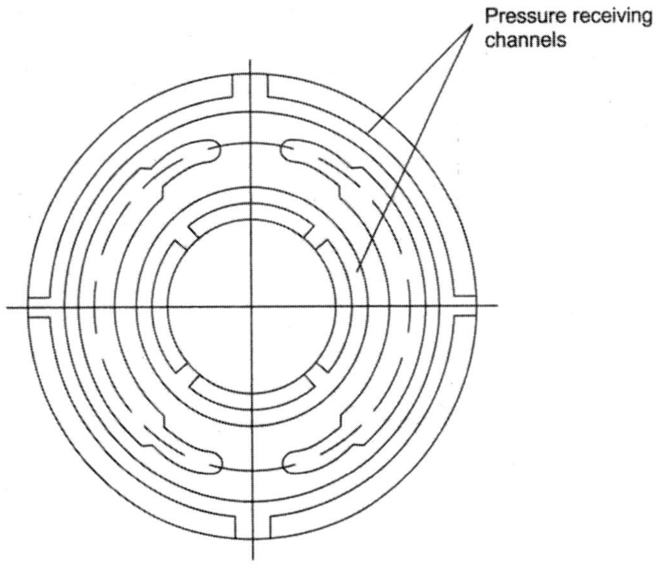

Figure 3.14. Rotating distribution disc

The distribution disc is fixed eccentrically with disc with respect the shaft center. During rotation, this disc is shifted with respect to the block. Each point of the disc rotates and in the form of the circle radius of which is equal to eccentricity '*e*' in (Fig. 3.14) the axis *y*-*y* relative to the portion of the inclined disc.

If the diameter of the circle M, which is located in the holes of the block which is having a diameter D_c, then the diameter of the inner and outer circle of the distribution disc must be made equal such that (as shown in Fig. 3.15) the circumference of the hole is at a distance of eccentricity '*e*' with respect to the outer diameter of the block for normal working of the unit.

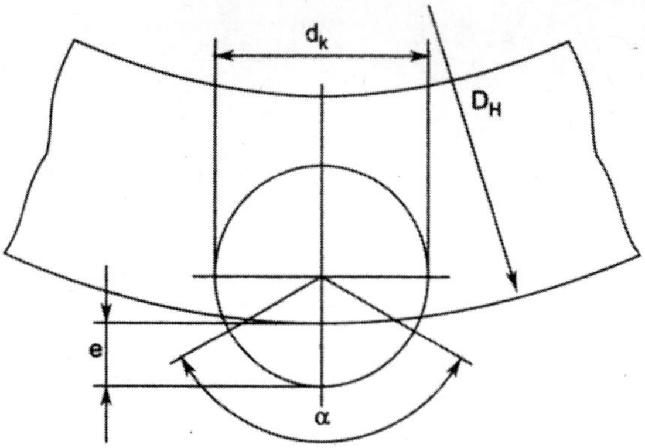

Figure 3.15. Scheme indicating the location of hole in the block with respect to the distribution disc

$$D_0 = D_c + d_k$$

and $$D_i = D_c - d_k$$

Where D_0 and D_i are the external and internal diameter of the segment of the distribution disc.

Since the distribution disc is located eccentrically with respect to the shaft centre and since the inclined disc is rotated with respect to the shaft centre, during rotation, the distribution disc will be connected to the hole one after another *i.e.*, to suction and to the delivery holes of the pumps.

The liquid coming out of the cylinder, when in contact with the surface, is determined by the external circle of the distribution disc and the casing of the pump. The space between the internal circle of the disc becomes the suction surface of the pump. When the rotation is reversed it will be delivery area.

In order to find the eccentricity '*e*' Fig. 3.14 may be referred. Maximum opening available from the hole, with the help of which the cylinder is connected with the suction surface or delivery surface, is determined as the eccentricity '*e*'. This area is limited by the diameter D_c and D_i and can be calculated if the area described by the outer diameter D_0 is changed as

$$A_i = \frac{\pi d_i^2}{4} \cdot \frac{\alpha}{2\pi} - \frac{d_k}{2}\left(\frac{d_k}{2} - e\right)\sin\frac{\alpha}{2} \qquad ...(3.27)$$

Where $$\alpha = 2 \text{ arc sin}\left(\frac{d_k - 2e}{d_k}\right)$$

Maximum instantaneous flow rate will be where $\phi = 90°$

Then $$q_{\Phi\max} = \frac{\pi d^2}{4}\,\omega R \sin\gamma \qquad ...(3.28)$$

Since $$A_i\, v_i = q_{\Phi\max}$$

$$A_i = \frac{\pi d^2\, \omega R \sin\gamma}{4 v_i} \qquad ...(3.29)$$

The velocity v_i in the hole is decided by D_0 and D_i

Normally v_i in the hole will be 6 to 8 m/sec. The eccentricity is determined by grapho-analytical method by using equation 3.27, 3.28 and 3.29.

3.6 PUMPS WITH ROTATING AND NON-ROTATING CYLINDRICAL BLOCK

The piston is in contact with the inclined disc through a ball in piston engaging in the circular groove in the inclined disc. Owing to the difficulty in construction and cost, piston is simply connected to the inclined disc, (Fig. 3.7). The return travel of the piston is done by rotating the cylinder or by rotating the disc. In the first we get axial piston pump and other elements rotating along with the cylindrical block, and nonmoving inclined disc. In the second axial piston pumps or its elements with nonmoving cylindrical block and rotating inclined disc.

Where constant contact at all times between the piston and inclined disc is necessary, springs are used inside the piston and having contact with cylinder (Fig. 3.7) sometimes a special arrangement is also made to get constant contact. For return of the piston has to be ensured for positive contact.

A safety valve is attached instead of spring. In that case flow rate will be higher than the normal flow rate by the amount of volumetric losses in the system.Connected with the delivery surface of the safety valve with the suction surface of the pump. In Fig. 3.15, scheme of axial piston pump with safety valve is shown. With the help of gear pump 2, the liquid flows from the suction reservoir 1, to the suction pipe or to the suction surface.

In regulated pumps the suction and delivery surfaces are inter changeable. That's why the entry of liquid is obtained either to suction surface or to the delivery surface. On the delivery pipe non-returns valve, 4 and 5 and attached to ensure unidirectional flow. To ensure the pressure stability at the delivery on the delivery line of the safety valve, a unidirectional pressure valve 3 is located.

The pressure developed by the safety valve must be sufficient enough to over come friction in the piston on the entire surface.

The friction between the piston and the cylinder in relative motion is equal to

$$F_{in} = m\, a_0$$

where, F_{in} – Inertia force

m – mass of piston

a_0 is the acceleration of the piston in relative motion is determined by the formula 3.9.

This friction force in relative motion tries to separate the piston from the inclined plate and is maximum when $\phi = 180°$ *i.e.*, during I[st] half of the suction, that's why the pressure of the safety valve should be determined for this condition. The pressure of safety valve should be

$$P_s = (m\,\omega^2 R \sin\gamma + F_{fri})\,\frac{4}{\pi d^2}$$

when P_s is the safety valve pressure

F_{fri} fractional force due to piston

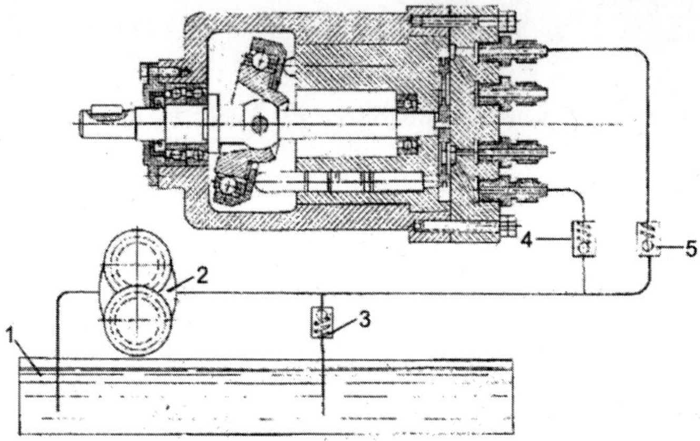

Figure 3.16. Axial Piston Pump with Controlled safety valve, Nonreturn valve and Gear pump

3.7 PUMP CONSTRUCTION

In Fig. 3.17 an axial piston pump, where cylindrical block is non moving is illustrated. On the shaft, the inclined disc 5 is located which during rotation, the piston 4 are reciprocated inside the cylinder 3, which is in non-moving block 2. The piston is always in contact with inclined plate due to the liquid pressure. In order to reduce friction angular contact bearings are fitted at the swash plate.

The movement of the piston is a simple harmonic motion. The pendulum part in this pump is played by the disc 7, which is rotating between the space of the surface of the block 2 and casing 1 of the distributed unit. The pendulum has circular canal with move number of holes 7. On the surface of the distributed unit is situated constantly canals 8 and 9. One acts as a delivery and the other for the leakage depending upon the position of the pendulum each line in cylinder block is either connected to delivery or suction area. The pendulum moves with the shaft helps of the curved shaped disc 6 (in figure the axis of the curved shaped disc is projected on the shaft axis) so that, the acting of pendulum coincides with the piston movement in the cylinder which is essential because the phase of work of pendulum leads by 90° to the inclined disc.

The rotation of the hub by the angle γ, lead to regulated unit, which is used for constant flow rate.

The entry and exit of liquid into and out of the pump is done by the distributed disc. The width of the canal or the window in distributed disc with rotating block must be equal to the diameter of the cylindrical hole with the help of which, the cylinder gets contact with suction or delivery surface. The diameter of the hole is selected based in the condition 2 to 3 m/s for self suction pumps and 3 to 4 m/s for non self suction pump.

In pumps of rotating distribution disc, the diameter of the hole in cylinder is adopted in the same manner as that in rotating block and non moving distribution disc but the area of opening window and the eccentricity e value are determined as per 3.27, 3.28, 3.29.

Material selection and the manufacturing methods are very important for the axial piston pump. The piston hole in cylindrical block must be made by grinding or by honing process. Before start the piston is lubricated with the pumping liquid but the hole by liquid movement. During side distribution of the liquid, the surface in contact with the distribution disc and the cylinder blocks are wet with the liquid before start. The surface of contact by the block with distribution disc must be perpendicular maximum allowable deviation 0.05mm for a length or 50mm. Mostly the block is made out of bronze; piston from chrome steel. After heat treatment the piston hardness must be HRC – 50 to 60. The distribution disc 5 made from chrome steal and heat treated to harness HRC – 58 to 62 the disc immediate to which the piston get compression force on moving and the casing of the inclined disc is made from bearing steel.

3.8 THE ADVANTAGES OF AXIAL PISTON PUMPS

The advantages of axial piston pumps are:

1. High efficiency when compared to radial piston pump due to smaller clearance for leakage.
2. Less weight.
3. For pressures upto 300 to 400 bar.
4. But the pumping oil must be pure hence needs a oil filter. The dust entering the space between the distribution disc and cylinder block, may cause disturbance in geometry.
5. Flow rate from 65 to 20,000 CC/sec at 60 to 200bar efficiency is 85 to 90% and in some cases upto 95%. The unit density of pump should be 2 to 3 N/kW. Axial pump with rotating inclined disc and non rotating cylindrical block are used for high velocity of rotation (n = 40 to 65 rev/sec) since during this the inertia force acting on the pump is minimum than the axial piston pump with rotating block.

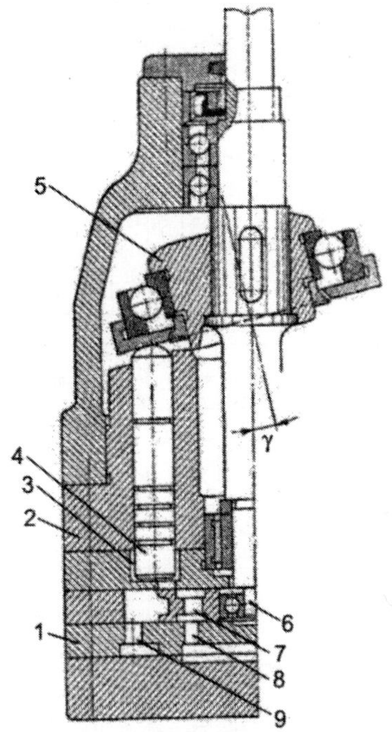

Figure 3.17. Axial piston pump with nonmoving disc

3.9 CHARACTERISTICS OF PISTON PUMPS

1. In Fig. 3.18 the characteristics are given as overall efficiency η vs delivery pressure *p*. This characteristic is drawn for the angle of inclined plate $\gamma = 10°$, constant flow rate and the pressure of auxilliary pump 4 bar.
2. In Fig. 3.19 the characteristics of axial piston pump is efficiency η vs speed *n* and indirection of the swash plate γ. This is called as universal characteristics because it gives the readings and results for a wide range of working of pump.
3. In Fig. 3.20 the characteristics of the radial piston pumps with 5 plungers is given as efficiency η and pressure *p* vs torque M. It must be mentioned that the characteristics of radial piston pump and axial piston pump are similar and analogy.

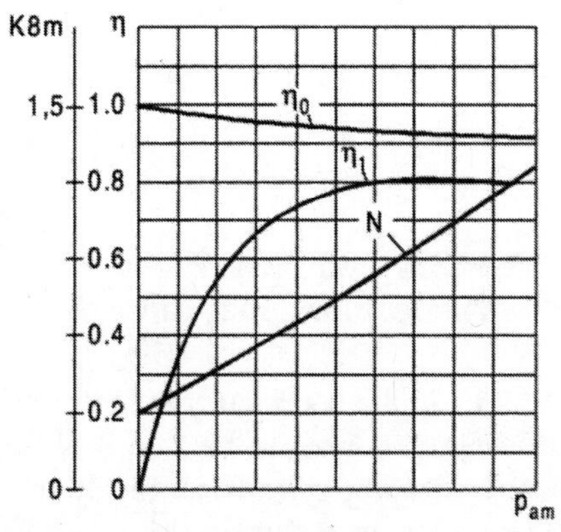

Figure 3.18.

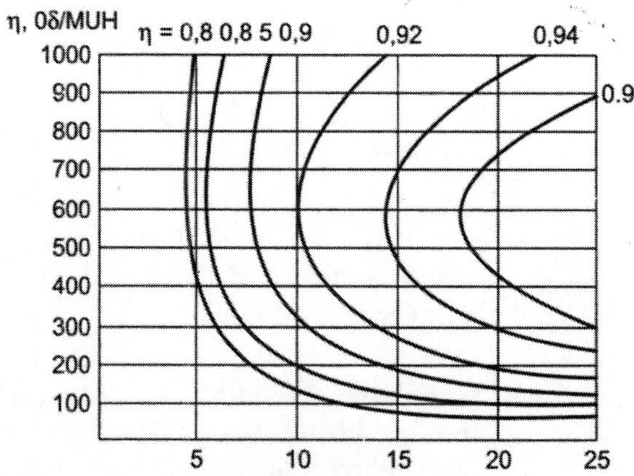

Figure 3.19.

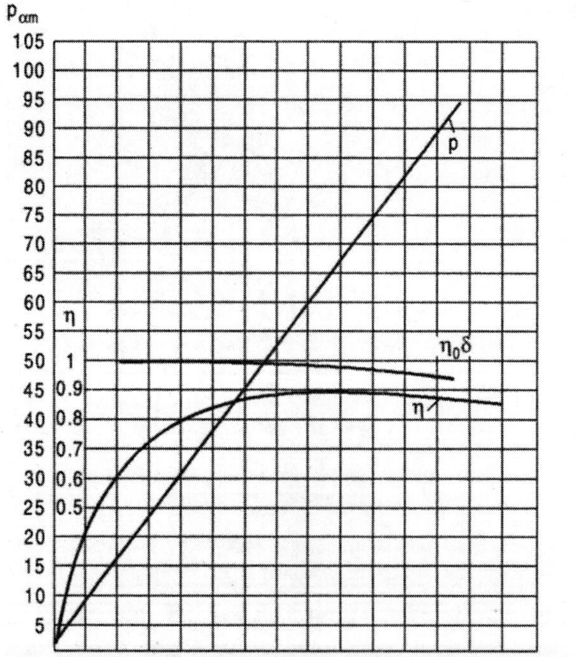

Figure 3.20.

CHAPTER 4

VANE PUMP

4.1 OPERATING PRINCIPLES AND APPLICATION

Rotary vane pumps are different from radial piston pumps in construction, although the operating principle is same.

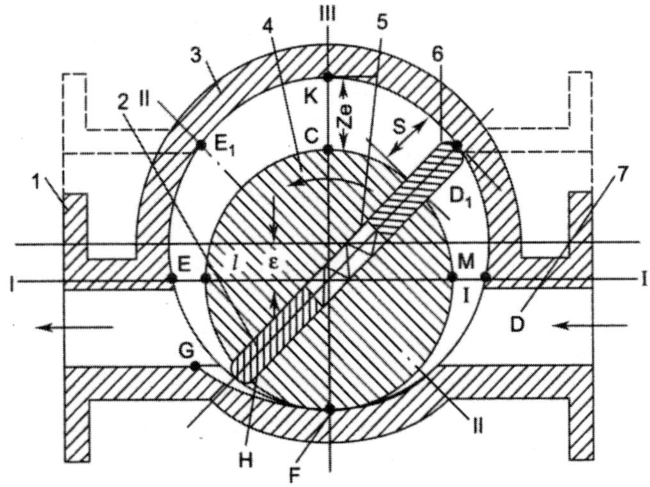

Figure 4.1. Schematic diagram of a vane pump

A schematic sketch of a rotary vane pump is illustrated in Fig. 4.1. The rotor 4, with two plates, called vanes, 2 and 6, are kept at 180° apart. It is eccentrically mounted on a shaft, with respect to the casing centre of the pump. A spring 5 is located between vanes. It is compressed by the vanes, and makes the vanes, to be in constant touch with the casing 3. The casing surface is designed, in phase with the contour of the vane movement. Both sides of the casing are closed by the cover plates, where the bearings are located. The starting point of the delivery pipe, at the lower part, is widened as shown by the line GF, in order to obtain compression for the fluid at the surface H.

During the rotation of the rotor as indicated in the figure 4.1, the vane 6 pushes the liquid from the casing to the delivery pipe 1 at the left side, as similar to that of a piston in rotary or axial piston pumps.

At the same time, the right side surface provides suction in the pump chamber. This process of work of the vane 6, starts at point D and ends at the point E. Further movement of the vane on the surface EFD will be in active. Like that, in one revolution of the vane, liquid is pushed from suction side to the delivery side, a volume equal to the area in casing DKELCMD. In the same way vane 2, follows the vane 6 and pumps simultaneously the fluid during the rotation of the rotor.

In Fig 4.1, the line sketch of a vane pump with two vanes is illustrated. During rotation of the rotor, the space after the vane is filled by the liquid, coming from the suction port. This liquid is pushed by the vane, there by the pressure is increased. This high pressure liquid goes out through the delivery port.

Due to movement of the blade, the delivery pressure prevails at the forward side of the blade, where as suction pressure exists at the other side of the blade, Although the blade is in constant touch with the casing, leakage takes place through the running clearance between casing and the blade. When one blade is under pressure, the other blade will be relieved from the high pressure during rotation of the blades. Thus each blade works under high pressure and relieved from the pressure during one rotation.

Minimum flow occurs when the blade enters or starts working. The work increase when the blade moves further. Maximum flow occurs when the blade reaches its vertical position. During further movement from vertical position, the flow starts reducing and attains minimum. The blade is relieved from the work.

The kinematic scheme of rotary vane pump is same as that of rotary piston pump. The flow, here also, is unsteady. In the piston pump, increasing the number of pistons reduces the fluctuation in flow. But the same is not always obtained, when the number of vanes are increased along the axis I-I, II-II, III-III etc. Increasing the number of vanes does not influence on the character of the movement of the fluid and its kinematic pattern.

Pulsation in flow is reduced, when number of blades are more. 4 to 12 number of blades are normally selected.

In Fig. 4.2 six vanes are adopted. In order to find the duration of contact of delivery port with suction port, lengths 1–2 and 3–4 are considered. The length 1–2 is situated and selected in such a way, that one blade enters the area 1, while other vane leaves the point 2. In order to get compression, the lower length 3–4 is made shorter than the length 1–2.

The flow rate is determined, from the height of blade exposed to the liquid in the casing during the period, when the blade passes from 1 to 2. In pumps having more number of blades the working part of the blade, exposed to the liquid during the period, when blade passes from 1 to 2, is small. The pulsation in flow is also small. During each revolution, each vane receives the liquid at one time and delivers the same quantity one time only. These machines are called single acting units.

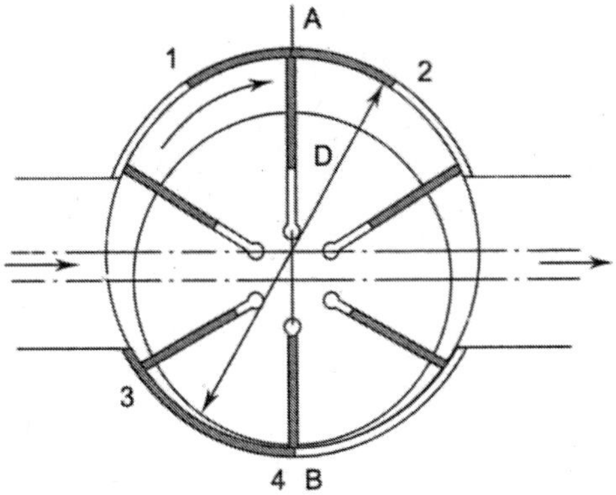

Figure 4.2. Pump with six vanes (single acting)

Increase in the number of vanes reduces the quantity of flow, but reduce the sealing effect considerably due to different clearances available between casing and the vane at each blade. For high pressure pumps, in order to reduce the leakage flow, from delivery to suction, number of blades is increased. For reducing non uniformity in flow, the stroke length **S** should be smaller *i.e.*, the suction starts from point D_1 instead of at D (Fig. 4.1). The reduction in area DKE to D_1KE, may reduce the unsteady character in flow. In order to get perfect sealing, more than 4 vanes, instead of two vanes are used.

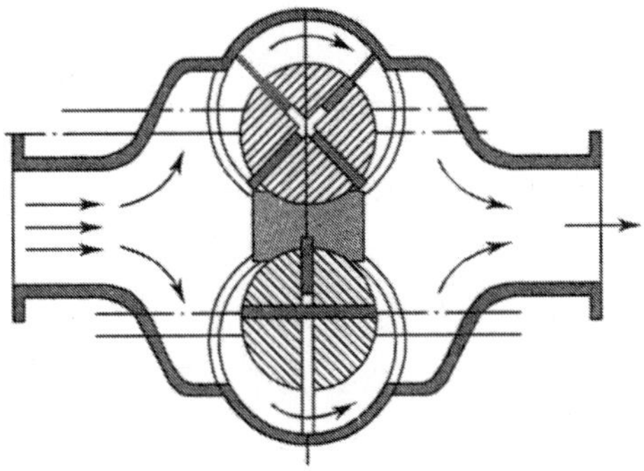

Figure 4.3. Two rotor pump with 4 blades in the same casing

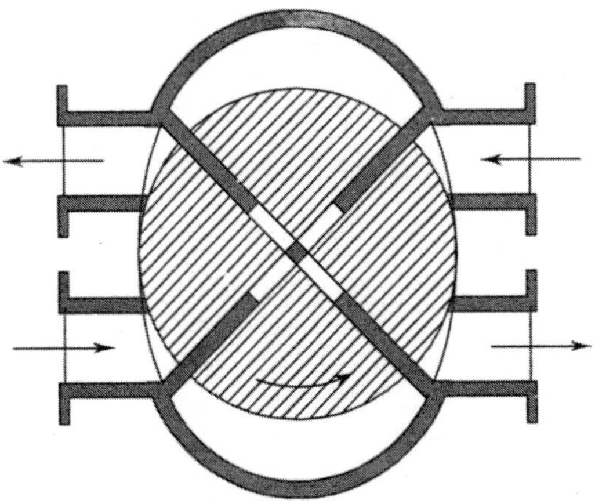

Figure 4.4. Double acting single rotor with 4 blades inside one casing

Only the reduction in the length of the working area in the casing DKE and increase in the number of vanes, can reduce the non uniform character in the flow. In Fig. 4.3 and 4.4, two different types of construction for a radial vane pump is illustrated. Vanes are made from quality material to possess low weight, and high elastic property, by which reduced dimensions are adopted. Due to positive contact between casing and the vane, the pump can work for a longer period, and even at higher suction heads. The disadvantage of this pump is the low mechanical efficiency due to constant rubbing of vane with casing. In order to reduce this friction and wear of the material, the vanes are fitted with rollers at the tip of the vane where the vanes are in contact with the casing.

Vane pumps are used for pumping different liquids especially viscous liquids. The flow rate Q ranges from 0.5 to 50 m^3/hour. Pressure head H ranges from 20 to 700 MWC. The speed n ranges from 300 to 3000 rpm and efficiency η ranges from 45% to 80%. These pumps are used for pumping lubricating oil, where low flow rate and high pressure are needed, which, other pumps cannot provide.

4.2 THE CHARACTERISTICS AND FLOW RATE OF VANE PUMP

Flow rate depends upon the kinematic movement of the vane system, inside the casing. The vane movement in the vane pump and the piston movement in the piston pump are schematically same. In Fig. 4.5, a kinematic scheme of the vane pump is illustrated. O is this centre of the axis of the vane, along which the vane moves. During the rotation of the rotor at an angle of ϕ, the movement or stroke of the vane $S = M_1D_1 = 2e$, where e is the eccentricity. Under a non separated flow condition, in the flow passage $DD_1E_1ELL_1M_1M$. Instantaneous flow in the passage M_1D_1 will be

$$q_{th} = \omega \times b \times S \times \rho = \omega \times b \times S \times \left[r + \frac{S}{2}\right] \quad \text{...(4.1)}$$

where q_{th} is the instantaneous flow and A is the vane area or actual flow area in the channel $= b \times S$.

b is vane width and

S is the stroke length $S = 2e$

u_{av} is the average peripheral velocity of the blade $= \omega\rho$

'ω' is the rotor angular velocity

$\rho = r + \frac{S}{2}$ is the radius of the point on the vane, where the velocity is u_{av}.

r is the radius of the rotor

Since, $S = 2e$, the equation (4.1) can be rewritten as

$$q_{th} = \omega b\left[(-r - e\cos\phi) + \sqrt{r^2 + e^2\cos^2\phi + 2er}\right]$$
$$\times\left[r + \frac{1}{2}(-r - e\cos\phi) + \sqrt{r^2 + e^2\cos^2\phi + 2er}\right] \quad \text{...(4.2)}$$

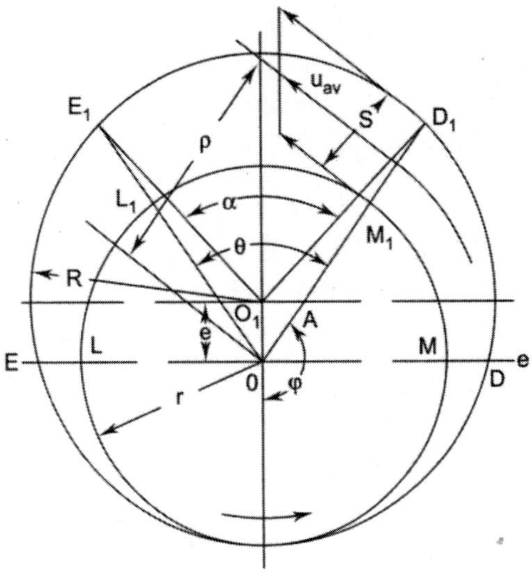

Figure 4.5. Kinematic diagram for the blade movement in a vane pump

The presence of cos ϕ indicates the non uniform motion of the fluid. When $\phi = 0$, $q_{th} = 0$ and when $\phi = 180°$

$$q_{th\,max} = \omega b2e\,(r+e) = 2\omega be\rho \qquad ...(4.3)$$

Equation (4.2) gives the instantaneous flow rate at any time, wherein only one vane is present at both suction and at delivery passages. Average theoretical flow Q'_{th} from Fig. (4.5) during one rotation of the rotor will be the area A = M_1, D_1 E_1 L_1 M_1 and Q''_{th} = A.*b*. For '*n*' revolution and Z, number of vanes.

$$Q'_{th} = Q''_{th}\ Zn = bAZn \qquad ...(4.4)$$

Taking Q_V as the volume occupied by one vane due to its thickness, A as the area occupied by one vane, and Z_1 as the number of vanes located in the considered area which is always less than the total number of vanes fitted in the rotor.

$$Q_{th} = \left(Q''_{th} - Q_v\right) Zn = bAZn - bAZZ_1n \qquad ...(4.5)$$

The number of blades Z for the calculation of flow rate is equal to the length in the casing D_1E_1 (Fig. 4.5), corresponding to the angle 'θ' *i.e.*, $Z = \dfrac{360°}{\theta°}$. Referring to Fig. (4.5),

A = area $O_1E_1D_1O_1$ + 2×area OD_1O_1 – area OL_1M_1O

$$\left.\begin{aligned} \text{area } O_1E_1D_1O_1 &= \frac{\alpha°\,\pi}{360°}.R^2 \\ \text{area } OD_1O_1 &= \frac{1}{2}\ O_1\,AOD_1 = \frac{1}{2}\ e\sin\phi\,(r+S) \\ \text{area } OL_1M_1O &= \frac{\theta°\,\pi}{360°}.r^2 = \frac{\pi r^2}{Z} \end{aligned}\right\} \qquad ...(4.6)$$

Substituting the values in the equation

$$A = \frac{\alpha^\circ \pi}{360^\circ} . R^2 + e \sin \phi (r + S) - \frac{\pi r^2}{Z} \quad ...(4.7)$$

Substituting the values in the equation (4.5)

$$Q_{th} = \frac{\alpha^\circ \pi}{360^\circ} . R^2 + e \sin \phi (r + S) - \frac{\pi r^2}{Z} - AZ_1 \quad ...(4.8)$$

The above equation is used to determine the basic dimensions of the pump. The characteristic pattern of these vane pumps are similar to that of reciprocating, gear or in general all positive displacement pumps *i.e.*, total pressure or total head (delivery head) and flow rate are independent. The reduction in flow rate occurs when total pressure is increased due to increasing leakage flow through the clearance between rotor and stator.

4.2.1 Flow Rate in a Single Acting Vane Pump

Referring to Fig. 4.6, a two bladed single acting vane pump is illustrated. From the triangle OO_1D, we get,

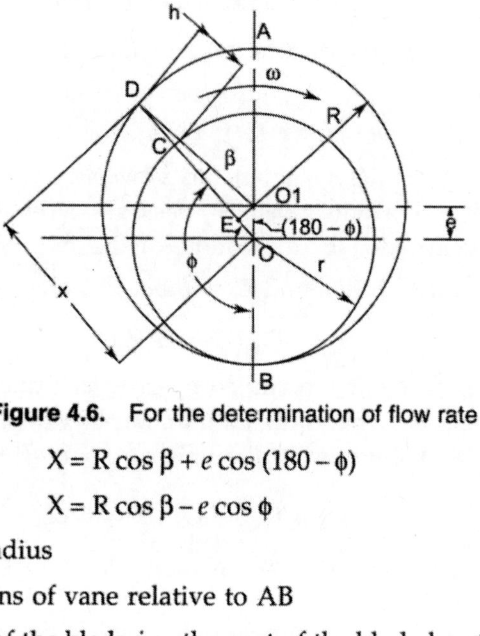

Figure 4.6. For the determination of flow rate

$$X = R \cos \beta + e \cos (180 - \phi)$$

or

$$X = R \cos \beta - e \cos \phi \quad ...(4.9)$$

where R is the stator radius

ϕ is the angle of rotations of vane relative to AB

The working height of the blade *i.e.*, the part of the blade length exposed to liquid in the delivery side of the pump

$$h = X - (R - e) = X - r \quad ...(4.10)$$

where $(R - e) = r$ is the rotor radius

Combining the equation (4.9) and equation (4.10)

$$h = R (\cos \beta - 1) + e (1 - \cos \phi) \quad ...(4.11)$$

Taking the angle β as almost zero $\cos \beta \approx 1$, then

$$h = e (1 - \cos \phi) \quad ...(4.12)$$

Differentiating equation (4.12) with respect to dt. The velocity of the blade v_b relative to the rotor in radial direction is equal to $\frac{dh}{dt}$.

$$v_b = \frac{dh}{dt} = e\omega \sin \phi \qquad ...(4.13)$$

The linear velocity of rotation of the blade *i.e.*, the rotation velocity of the gravity centre of the working area of the blade point C in Fig. (4.6) is equal to the differentiation of angular velocity of rotation of the rotor on the distance from the axis of rotation of rotor upto the point C.

Assuming, $\beta \approx 0$ and $\cos \beta \approx 1$

From Fig. (4.6)

$$\begin{aligned} OC &= X - 0.5h \\ &= R \cos \beta - e \cos \phi - 0.5\, e\, (1 - \cos \phi) \\ &= R - 0.5\, e\, (1 + \cos \phi) \end{aligned} \qquad ...(4.14)$$

The rotation velocity of C will be

$$V_{ro} = (X - 0.5h)\, \omega = [R - 0.5e\, (1 + \cos \phi)]\, \omega \qquad ...(4.15)$$

Neglecting the blade thickness, instantaneous flow rate is equal to the product of velocity V_{ro} and the area of working side of the blade. Taking breadth of the pump as 'b'

$$Q'_\phi = V_{ro}\, bh = [R - 0.5e\, (1 + \cos \phi)]\, \omega be\, (1 - \cos \phi) \qquad ...(4.16)$$

From equation (4.16) it is evident that when ϕ = 90°, instantaneous flow rate is minimum and is equal to

$$Q'_{\phi min} = (R - 0.5e)\, \omega be \qquad ...(4.17)$$

since, 0.5e is very small value, when compared to the stator radius R, it can be neglected. Equation (4.17) can be written as

$$Q'_{\phi max} = R\omega be \qquad ...(4.18)$$

When ϕ = 180° flow rate will be maximum and is equal to

$$Q'_{\phi max} = 2\, R\omega be \qquad ...(4.19)$$

Maximum flow rate of the pump is independent of number of blades as per the equation (4.18) and if there are 2 blades, the flow rate will be twice if 0.5e and blade thickness is neglected.

For Z number of blades minimum flow rate will be when the blade starts working *i.e.*, entering the space 1 in the line 1–2. At that moment, angle

$$\phi = \left(\pi - \frac{a}{2}\right) \text{ or } \left(\pi - \frac{\pi}{Z}\right), \text{ where } a = \frac{2\pi}{Z}.$$

From equation (4.16)

$$Q'_{\phi min} = \left[R - 0.5e\left(1 - \cos \frac{\pi}{Z}\right)\right] \omega be \left(1 + \cos \frac{\pi}{Z}\right) \qquad ...(4.20)$$

For example, taking number of blades Z = 6, $\frac{\pi}{z} - \frac{\pi}{6} = 30°$

and $\cos 30° = \frac{\sqrt{3}}{2}$, then equation (4.12) will be

$$Q'_{\phi min} = \left[R - 0.5e\left(1 - \frac{\sqrt{3}}{2}\right)\right] \omega be\left(1 + \frac{\sqrt{3}}{2}\right)$$

After reconstructing $Q'_{\phi min} = 1.87\ \omega be\ (R - 0.067e) \approx 1.87\ \omega be$

However the maximum flow rate will be same as $Q'_{\phi min} = 2R\ \omega be$.

From the above, it is seen that for a 6 bladed pump, the ratio of maximum flow rate to minimum is 1.07 *i.e.*, very near to '1'. From this, it is evident that increasing number of blades reduces the pulsation in flow and the flow is steady and the considering blade effect thickness. The instantaneous flow rate δ, reduces due to vane thickness.

The velocity of vertical movement relative to rotor is $v_b = e\ \omega \sin\phi$. Correspondingly the flow in the space due to the reduction in the height of the working length of the blade is

$$Q'_0 = v_b\, b\, \delta = e\ \omega b\ \delta \sin\phi \qquad ...(4.21)$$

Considering the thickness, the instantaneous flow of 2 vane pump will be

$$Q_\phi = Q'_\phi - Q_0 = [R - 0.5e\ (1 + \cos\phi)]\ \omega be\ (1 - \cos\phi) - e\omega b\ \delta \sin\phi \qquad ...(4.22)$$

When $\phi = 0$

$$Q_{\phi min} = (R - 0.5e)\ \omega be - e\omega b\ \delta\ (R - 0.5e - \delta)$$

$$= ebn\ (2\pi R - \pi(e - 2\delta)) \qquad ...(4.23)$$

Maximum flow rate will be same

$$Q_{\phi max} = 2R\ \omega be$$

For Z number of blades the instantaneous flow rate is determined when the flow coming out to fulfill the space, released by the blade, which increases the load in rotor area in the delivery space. Taking 'α' as the angle between the blades.

$$Q_\phi = [R - 0.5e\ (1 + \cos\phi)]\ \omega be\ (1 - \cos\phi) - e\omega b\ \delta \times [\sin\phi + \sin(\phi + \alpha) + ... + \sin\left[\phi + \left(\frac{Z}{2} - 1\right)\alpha\right] \qquad ...(4.24)$$

The average flow rate for one revolution

$$q = Z\,e\omega b \int_{\left(\pi - \frac{\alpha}{2}\right)}^{\left(\pi + \frac{\alpha}{2}\right)} \left[R - 0.5e\,(1 + \cos\phi)\right](1 - \cos\phi)\frac{d\phi}{\omega} - 2eb\ \delta z \qquad ...(4.25)$$

The IInd part of the equation (4.25) indicates the flow due to the free area obtained in the delivery space due to vane loading in the rotor area during one revolution.

After integrating the equation (4.25) will be

$$q = eb\left(R\alpha + 2R\alpha \sin\frac{\alpha}{2} - \frac{e}{4}\alpha + \frac{e}{4}\sin\alpha - 2\delta\right) \qquad ...(4.26)$$

If Z, the no: of blades are more, then

$$\sin\frac{\alpha}{2} \approx \frac{\alpha}{2} \text{ and } \sin\alpha = \alpha$$

Then, $q = Zeb\,(2r\alpha - 2\delta)$ ...(4.27)

Since, $Z\alpha = 2\pi$

$A = (Zeb - \delta Z)$ ...(4.28)

For a speed of n rpm

$$Q_0 = q \cdot \frac{n}{60} = 2eb(2\pi r - \delta Z)\frac{n}{60}\ \text{m}^3/\text{sec} \qquad ...(4.29)$$

4.3 FORCES ACTING ON THE PUMP

Taking the pressure difference acting on the blade is p, the force F acting at the centre of gravity of the working area of the blade is equal to the area of the working blade multiplied by the pressure p.

The moment of the force F about the rotor axis is

$$M = (X - 0.5\,r)\,pbh \qquad ...(4.30)$$

Substituting the value of X and h

$$M = \text{peb}\,[R - 0.5e\,(1 + \cos\phi)]\,(1 - \cos\phi) \qquad ...(4.31)$$

From equation (4.31) it is found that the torque also changes as that of flow rate during one revolution of the blade. At 180°, the moment is maximum.

$$M = 2\Omega peb \qquad ...(4.32)$$

It is necessary to have a firm contact between blade and the stator. This ensures that there should be an inertia force which is equal to the centrifugal force acting on the stator by the blade. In relative movement this centrifugal force of inertia depends upon the angle of rotation of the blade. The coroliss component of the force of inertia increases the friction of the blade, when it moves inside the rotor slot.

In order to get firm contact between blade end and the stator, liquid under delivery pressure is applied, at the bottom of the blade, in the rotor space available under the blade. The liquid from the delivery area will be flown to the bottom of the blade. Under this condition blade is acted upon by the liquid pressure at the bottom of the blade. Centrifugal force of inertia in relative flow and coroliss force of inertia are acting at present.

Centrifugal force of inertia is equal to the product of mass of the blade and the centrifugal acceleration is the distance from axis of rotation upto the centre of gravity of the blade.

It will be

$$R_s = X - \frac{l}{2} \qquad ...(4.33)$$

where l is the blade length

Considering equation (4.9) and taking $\beta \approx 0$ and

$$\cos \beta \approx 1$$

$$R_s = R - \frac{l}{2} - e \cos \phi \qquad ...(4.34)$$

The centrifugal acceleration is $a_n = \left(R - \frac{l}{2} - e \cos \phi\right)\omega^2$ ...(4.35)

Centrifugal force will be

$$F_{centri} = m\left(R - \frac{l}{2} - e \cos \phi\right)\omega^2 \qquad ...(4.36)$$

where m is the mass of the blade.

Inertia force in relative flow will be the product of mass of the blade and its acceleration. Since the rotation movement is a return and linear, the acceleration of the blade in relative movement will be in the IInd order with respect to the relative shift of the blade during the time t.

$$a_0 = \frac{d^2h}{dt^2} = e\omega^2 \cos \phi \qquad ...(4.37)$$

The inertia force in relative movement is

$$F_{inert} = -me\,\omega^2 \cos \phi \qquad ...(4.38)$$

Inertia force will be acting opposite to the acceleration. When ϕ changes from 0 to 90°. The inertia force in relative movement tends to remove the blade from the casing. When ϕ changes from 90° to 180° the blade is pressed to the stator.

Same thing happens when ϕ changes between 180° to 270°. When ϕ changes from 270° to 360°, inertia force tries to remove the blade from the casing. Maximum inertia force will be when ϕ = 180°. Then

$$F = me\omega^2 \qquad ...(4.39)$$

At this time the blade is pressed into the stator. Neglecting friction

$$N_{max} = pb\delta + Ma_{nmax} + Ma_{omax} \qquad ...(4.40)$$

Where, N_{max} – the force of blade on stator

a_{nmax} – Maximum value of centripetal acceleration

a_{omax} – Maximum acceleration of the blade in relative motion

Since, $a_{nmax} = (R + e - 0.51)\omega^2$

and $a_{omax} = e\omega^2$

$$N_{max} = pb\,\delta + m\,(R + 2e - 0.5l)\,\omega^2 \qquad ...(4.41)$$

Now, considering the coroliss, this component increases the friction between stator and blade and acting in the radial in the direction *i.e.*, normal to the relative motion.

$$F_{iner} = 2me\,\omega^2 \sin \phi \qquad ...(4.42)$$

From equation (4.34) the power N_{max} depends upon the blade thickness δ. Friction increases when blade thickness increases although N_{max} increases. It is found that this force is so much that even the blade cannot be moved. In order to reduce the friction force the thickness of the blade is taken as 2 to 2.5mm.

The unit pressure exerted on the blade on stator from the power N_{max} is

$$\sigma = \frac{N_{max}}{b\delta} \quad ...(4.43)$$

Taking the value of N_{max}

$$\sigma = p + Pl\,(R + 2e - 0.51)\,\omega^2 \quad ...(4.44)$$

where ρ is the density of the blade material.

Since, the area of contact between the stator and the blade is very small. When compared, the area of the blade side area the pressure at the point of contact enormously increases which causes the wear of the contact portion. The high temperature produced at the contact point also creates more wear at this point.

It is found that the bearings in these pumps are subjected to heavy load, due to the delivery pressure acting on one side of the rotor, which changes according to the position of the blade the maximum force will be

$$P = p2Rb \quad ...(4.45)$$

where, p – pressure difference between suction and delivery

R – Stator radius

b – Width of the casing

From equation (4.45) the load on the bearing can be estimated. In order to keep the bearing size to an acceptable dimension in the pump, the pumps are designed normally upto 20 to 30 bar. It should also be explained that by regulating the eccentricity, the flow can be changed. In this case, the pump is called as regulated pump otherwise called as non regulated pump.

4.4 PUMP WITH TWO SUCTIONS AND TWO DELIVERIES

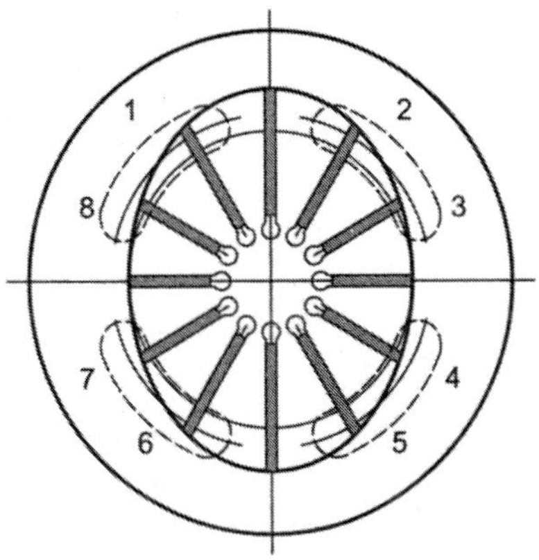

Figure 4.7. Pump with six vanes and with two suction and two deliveries

Pump with two suctions and two deliveries is shown in Fig. (4.7). When the rotor rotates in clock wise direction, 8–1 will be suction and area 2–3 will be delivery Liquid comes through the window located at the side is the delivery and suction port. The area 4–5 will be the IInd suction and area 6–7 will be the IInd delivery. Since, the delivery and suction are opposite to each other, the bearings are not loaded on one side as in the previous case *i.e.*, in pumps with single suction and single delivery.

Because of this, the pumps are operated for higher pressures. In single suction and delivery, unit pump delivery pressure is used to maintain firm contact between blade and casing. The stator inside surface is made into 8 parts 1–2, 3–4, 5–6, 7–8 where compression of the liquid takes place. The places 2–3, 4–5, 6–7, 8–1 are made in the curved from and is in between the previously 1–2, 3–4, 6–7 and 8–1 parts.

Neglecting the blade thickness and since there are two suctions and two deliveries, the flow rate will be

$$Q_1 = 2hbv \qquad \text{...(4.46)}$$

where h – Working height of the blade

b – The rotor width

V – The velocity of the centre of gravity of the working part of the blade working height $h = R_1 - r$

$$V = \frac{R_1 + r}{2}\omega$$

So,

$$Q_1 = \left(R^2 - r^2\right) b\omega \qquad \text{...(4.47)}$$

Where, R – stator radius at the part 1–2

r – rotor radius

If stator radius at 3–4 sections larger than rotor radius part of the liquid flows from delivery to suction. This flow will be

$$Q_2 = \left(R_2^2 - r^2\right) b\omega \qquad \text{...(4.48)}$$

where, R_2 is the radius of stator at section 3–4

The flow rate, without considering blade thickness, will be

$$Q' = (Q_1 - Q_2) = \left(R_1^2 - R_2^2\right) b\omega \qquad \text{...(4.49)}$$

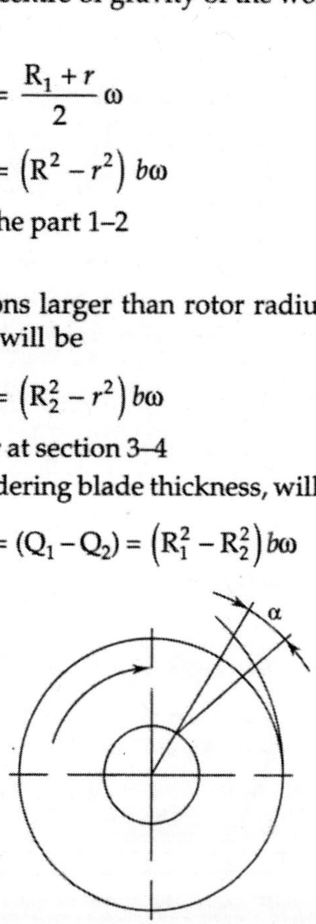

Figure 4.8. For determination of flow rate in two suction pumps

Actually the flow will be less than the above calculated theoretical flow due to blade loading, and the rotor construction.

With thickness of the blade δ the flow rate

$$Q_0 = \left(R_1^2 - R_2^2\right) b\omega - 2b\,\delta\, zn\,\frac{r_1 - r_2}{\cos\alpha} \quad \text{...(4.50)}$$

where α is the angle of inclination of the blade Fig. (4.8).

The flow is independent of angle of rotation of the rotor. In the same manner the moment M for a pressure difference of p will be

$$M = 2p\,\frac{R+r}{2}\left(R-r\right)b$$

$$= pb\left(R^2 - r^2\right) \quad \text{...(4.51)}$$

4.5 SELECTION OF PARAMETERS OF VANE PUMP

Normally (R – r) will be 0.4 from the total blade height l. Experiment indicates that increase of (R – r), the bending blade in the rotor space is more. Reduction in (R – r), reduces the flow rate and increases the pump weight. In some design the blade is positioned at angle to have strength. The direction will be 6 to 13° with respect to the radial line. Number of blades will be 6 to 12 and thickness 2 to 2.5 mm. The width b will be 20 to 30 mm rotor rotation will be 1000 to 2000 rev/min.

The velocity of flow in windows will be 1.5 to 2 m/sec. The space between windows in disc is 0.2 to 0.6 mm higher than the distance between two blades at the outer end for two phase entry pumps. Material selection is very important and its manufacturing process. High temperature resistant materials are prepared for blade since very high temperature is taking place at the contact points of the blade with stator. The hardness of the blade after heat treatment should be HRC 58 to 62. The inner surface of the stator must have HRC 56 to 60 and rotor after manufacturing heat treatment must have HRC 58 to 60. Bronze is used for side cover where as high strength steel is used for stator and rotor.

The suction velocity in the windows must be sufficient fill the space between two blades with in the given time.

Since the inertia force is directed from rotor centre to outwards it will reduce the delivery pressure of the liquid between two blades. This in turn decides the speed of the rotor.

Referring to Fig. (4.9) the dynamic stability of the pump taking into inertia for also account.

$$dp\, dA = \rho dA dr \omega^2 r \quad \text{...(4.52)}$$

When, dA is the area of the selected segment

ρ is the density

Integrating $$p = \rho\frac{\omega^2 r^2}{2} + c$$

When, $$r = r_{ro}, p_{ro} = \rho\frac{\omega^2 2^2_{ro}}{2} + c$$

So, $$p_{ro} = p - \rho\frac{\omega^2}{2}\left(r^2 - r_o^2\right) \quad ...(4.53)$$

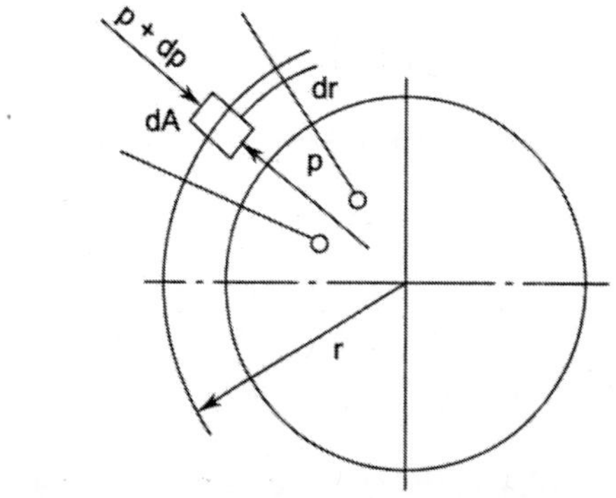

Figure 4.9. Determination of pressure during rotor rotation

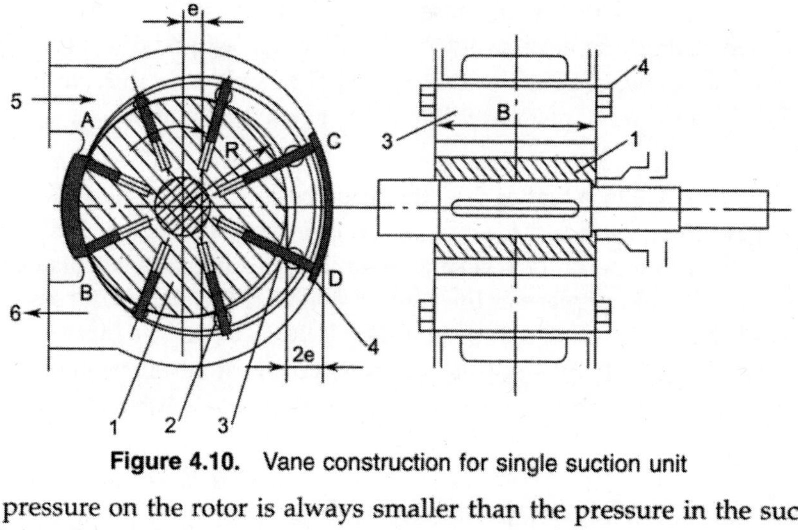

Figure 4.10. Vane construction for single suction unit

So the pressure on the rotor is always smaller than the pressure in the suction area. With the increases of the rotor space the process in blades reduces for normal working. The rotor pressure should more than the vapour pressure p_{vp}. The flow rate for single suction will be 5 to 150 lit/min and delivery pressure is 15 to 20 bar. For two suction units flow will be 5 – 200 lit/min up to 250 lit/min and the delivery pressure is 65 to 70 bar. Efficiency of the unit depends upon the manufacturing process and flow rate. It ranges from 50% to 70% and even up to 85%.

Chapter 5

Gear Pumps

5.1 INTRODUCTION

Gear pumps are widely used in industries, especially, where positive displacement of liquid is essential, such as pumping of oil from oil tank to the furnace, supply of lubricating oil to various lubricating points in a machine etc. The advantage of gear pump is compact, light weight, continuous flow, long life, and can be used for continuous operations. Self lubricated gear pumps are mostly used for a flow of 0.5 to 250 m^3 / hr and for a delivery pressure of 0.5 ata to 13 ata. The speed ranges from 400 to 5000 rpm. Pumps operated in this region give high efficiency. When the delivery head exceeds 13 ata, the volumetric efficiency goes down in simple gear pumps (Fig. 5.1). The volumetric efficiency in these pumps can be increased, by reducing the radial and axial clearances between the gears and the housing and also by providing proper surface finish for the contour of gear teeth. However this process increases the manufacturing cost. Gear pumps can work even up to 100 ata without any reduction in volumetric efficiency. Under positive suction head, these pumps work continuously and efficiently, and need a minimum regulation, when it runs at constant speed.

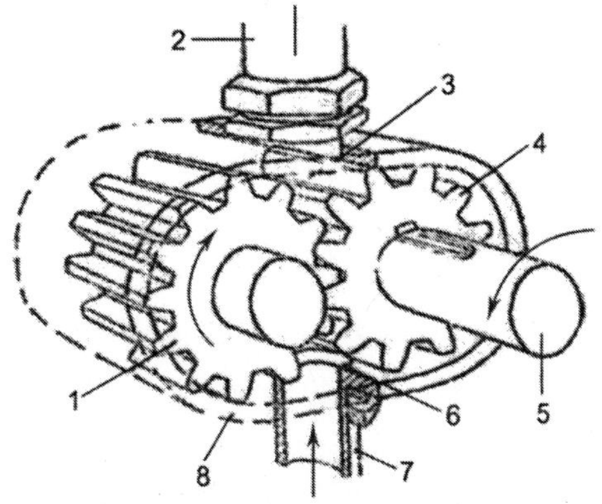

Figure 5.1. Spur gear pump

5.2 GEAR PUMP CONSTRUCTION

A simple sketch of a gear pump with two gears in mesh is given in figure 5.1. It consists of two gears, the driver (4) and the driven (1). The driver gear shaft (5) is connected to the prime mover through coupling. Due to the mesh between these two gears, the driven gear also rotates along with the driver, but in opposite direction. Both the gears are kept in a housing (8). The radial and axial clearances between the housing and the gears are kept minimum in order to avoid the leakage. The housing is connected to a suction pipe (7). The space (6) between gears, the housing and the suction pipe, called suction chamber, is filled with pumping fluid, coming from the suction pipe. When the pump runs, oil from the suction chamber (6) is taken by the rotating gears through the space between the gear teeth by both gears. This oil is collected at the space (3) called delivery chamber. Due to firm contact between the teeth of the driven and the driver, oil cannot return back to the suction chamber (6), instead it passes through the delivery pipe (2). The liquid occupies the space between two teeth of the gears and is pushed from the suction chamber 6 to delivery chamber 2, due to the rotation of the gears. The movement of liquid cannot be stopped whenever the pump runs. Thus, the gear pump is a positive displacement pump. The flow rate is the function of the area between two gear teeth and the speed of rotation. It is independent of delivery pressure.

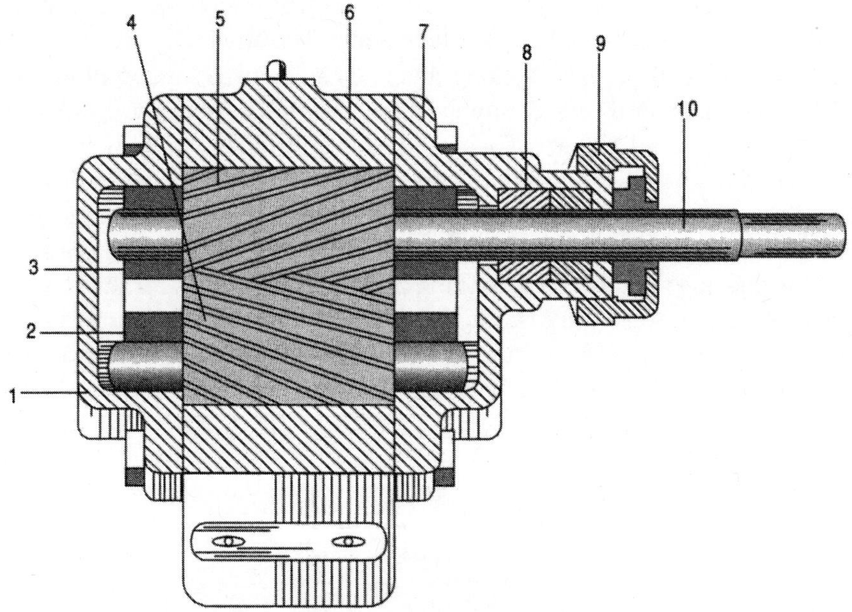

1. End cover. 2 Driven gear bearings. 3. Driver gear bearings. 4. Driven gear. 5. Driver gear. 6. Main cover. 7. End cover. 8. Sealing arrangement (mechanical seal or stuffing box). 9. Gland cover. 10. Gear shaft.

Figure 5.2. Gear pump with inclined teeth

The flow will be uniform so long as the speed is constant.

Although any form of tooth profile can be used for gear pumps but, to get positive contact without clearance between them, corrected involute profile is recommended and adopted due to the following advantages.

1. The error in the distance between the centers of the gears, does not affect the performance. This is very important, since wear will be more if the error is more.
2. The maintenance of gear pump is less, and the performance even after a long run remains unaltered. The discharge is uniform.
3. The manufacturing process is simple and
4. The repair and reconditioning is simple by which positive contact is achieved between driver and driven, which yields to higher volumetric efficiency.

5.3 ADVANTAGES AND APPLICATION OF GEAR PUMP

The main advantages of the gear pump are

- Considerably smaller and compact in size.
- Smaller weight for the given head and flow rate when compared to other pumps.
- Simple in construction and easy to manufacture. Minimum number of parts in pump.
- Dependable operation even at high speeds, since both the gears evenly rotate.
- Long life.
- Practically uniform flow rate without fluctuations.

These pumps are commonly used for viscous liquids such as furnace oil, hot viscous liquids in hydraulic circuits and in control systems.

5.4 LIQUID FILLING IN THE SPACE BETWEEN TEETH

The space between the two successive teeth of the driver and the driven must be fully filled with liquid, in order to avoid,

1. Reduction in the flow rate
2. Impulsive force created due to the pulsating radial force on the gear, due to the gas present in the liquid undergoing compression and decompression, in the compressed space, between teeth, during the rotation of the gears, and
3. Noise and vibration.

This depends upon:

1. The duration, at which the space between the gears is in contact with the suction chamber
2. Speed of rotation
3. The pressure at the suction chamber,
4. Form, shape and the dimensions of the teeth
5. The compression space between the teeth, of the driver and the driven gears
6. The volume of the vapour or gas present in the fluid, in the space.

The liquid in the suction chamber often contains vapour or gas. This liquid, on reaching the space between the teeth, gets compressed for a very short duration due to the rotation of the gears. Instantaneously a very high pressure is developed especially when the fluid reaches the compression space. This produces noise and vibration, very high radial force for a short duration of the casing, which produces fatigue stress on the shaft and on the bearings especially on the driver shaft and on the bearing, thereby reducing the life of the pump, reduction in flow rate due to increased radial and axial clearances. In order to reduce this hammer blow in the casing especially at the delivery side of the pump, a small hole is made at this space (Fig. 5.12 and 5.13), through which the liquid starts entering the space between the teeth, till the space pressure is equal to the delivery chamber pressure. This process raises the pressure gradually to the delivery pressure

Taking an elementary strip (Fig. 5.3) with a circumfrancial length $r\,d\phi$, radial length dr and width b at a radius r in the space between the teeth, the radial force acting on the elementary strip will be $p\,rd\phi\,b$, at the bottom side and $(p + dp)\,rd\phi$ at the top side.

The centrifugal force $F_c = \rho\,r\,d\phi\,dr\,b\,r\,\omega^2$

Resolved component of this force acting on the pump casing along the radial direction is

$$prd\phi b - (p + dp)\,rd\phi b + \rho rd\phi\,dr.b\omega^2 r = 0$$

or
$$dp = \rho\omega^2 rdr \qquad \text{...(5.1)}$$

Where p is the hydrodynamic pressure acting on the cylindrical section at radius 'r'

$(p + dp)$ is the hydrodynamic pressure acting at radius $(r + dr)$

ρ is the density of the liquid

ω is the angular velocity of the gear.

$d\phi$ is the angle subtended between this sections in the space considered

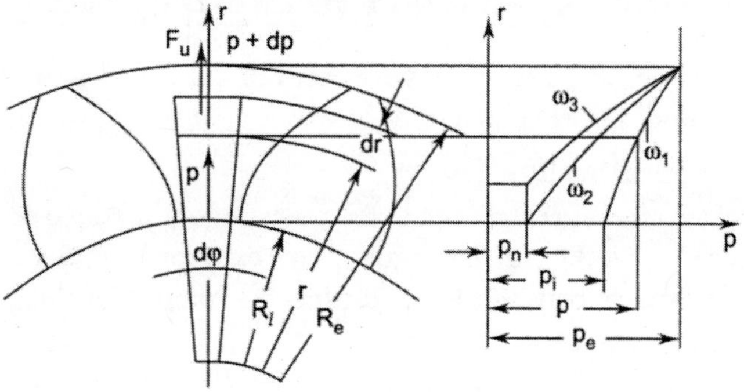

Figure 5.3. Calculation of flow passage between two gears

Integrating the above equation between r and R_e, the pressure

$$p = p_e - \Delta p_c = p_e - \rho\omega^2\,\frac{R_e^2 - r^2}{2} = p_e - \gamma\,\frac{u_e^2}{2g}\left(1 - \left(\frac{r}{R}\right)^2\right) \qquad \text{...(5.2)}$$

Where u_e is the peripheral velocity at radius R_e,

Δp_c is the pressure drop due to centrifugal force, and

p_e is the pressure at radius R_e.

The pressure p is a function of ω and r. Under constant speed ω, the pressure 'p' reduces, due to the centrifugal force, from the periphery towards centre in the form of the parabola. The pressure p_e at the outer circumference of the gear, when it comes in contact with the suction chamber will be equal to suction pressure p_s. When the speed ω is increased under constant pressure p_e, the pressure 'p_i'', in the space between gears also reduces.

The maximum allowable speed $u_{e\ max}$, $= u_{e\ allow}$, will be, when the pressure p_i reaches the vapour pressure $p_{vp}, p_e = p_{e\ min}$. Practically $p_{e\ min}$ is selected slightly above the vapour pressure in order to avoid vapour formation in the space between the teeth as well as in suction chamber. $p_{min} \geq p_{vp} = p_{vp} + 0.3$ ata. The allowable speed $u_{e\ max}$ will be

$$u_{e\ max} = u_{e\ allow} = \sqrt{\frac{2g\left(p_e - p_{vp} - 0.3\right)}{\gamma\left(1-\left(\frac{R_i}{R_e}\right)^2\right)}} \qquad ...(5.3)$$

and the pressure

$$p_e = p_{i\ min} + \gamma\frac{u_{e\ max}^2}{2g}\left(1-\left(\frac{R_i}{R_e}\right)^2\right) \qquad ...(5.4)$$

In order that the gear pump works better, the liquid should occupy entire width of the gear and should be in contact with the suction chamber for longer duration during the rotation, for an angle ϕ. For a velocity of u_e = 6 to 8 m/sec, angle ϕ will be 45° and above so that the space will be in contact with the suction chamber for a longer time. When viscosity increases, the angle ϕ increases in order to accommodate the low flow velocity of the fluid. The angle ϕ = 120 to 300° (figure 5.4). /16/

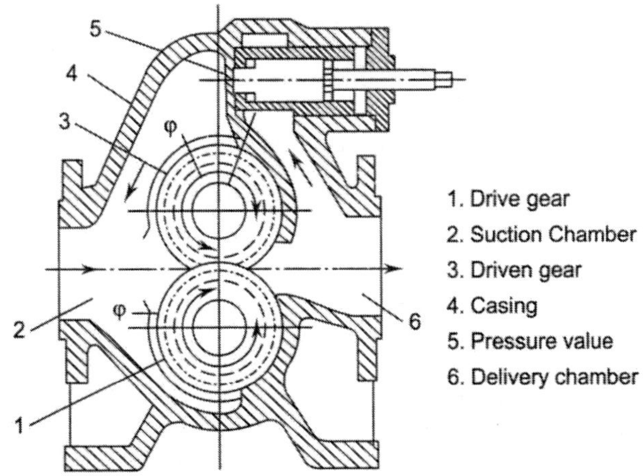

Figure 5.4. Space available for suction chamber (angle ϕ)

The velocity will be 1 to 2 m/sec at inlet and 1.5 to 3 m/sec at outlet depending upon the viscosity of the liquid and the delivery pressure. The actual space available between the driver and the driven gear teeth, called the compression area, depends upon the radial clearance between the space of root diameter of one gear and the outer diameter of the other gear as well as the side clearance between teeth, and also depends upon the presence of air or vapour in the liquid, which reduces the flow rate, especially at higher pressure. The compressed air at delivery side expands when it comes to suction side, which also occupies space and reduces the flow rate as well as the suction effect. Flow variation also takes place. This is more when viscosity increases. It is always recommended to reduce the compression area.

The following table gives the recommended velocity for liquids of different viscosities, in order that the liquid occupies the entire space available between teeth.

Table 5.1.

Kinematic viscosity η cs	12	45	75	155	305	530	760
Velocity v m/s	5	4	3.7	3	2.2	1.6	1.3

5.5 FLOW RATE AT ANY POSITION OF THE GEAR WHEELS

Instantaneous flow rate (q_{ins}) is the sum of the flow through driver (q_{1ins}) and the driven (q_{2ins}), *i.e.*, $q_{ins} = q_{1ins} + q_{2ins}$

However $$q_{ins} = \int_a u_n \, da$$

Where u_n is the resolved component of peripheral velocity u, along the normal n to the mixing area, and dA is the elementary area of the tooth surface. Integrating for the entire surface a between the limits of the top and the contact point, since $u = r_x\omega$ and $u_n = u \cos(nu)$

$$q_i = \int_a u_n \, dA \cos(nu)$$

The magnitude $dA \cos(nu)$ is the projected area of the plane passing through the point B and axis (Fig. 5.5*a*). Since u is a function of 'r_x' and remains constant along the entire width of the gear, the elementary area $dA = bd_x$

So,

$$q_i = \int_A u dA \cos(nu) = \int_{r_a}^{R_e} r_x \omega b \, dr_x = b\omega \frac{R_e^2 - r_a^2}{2} \quad ...(5.5)$$

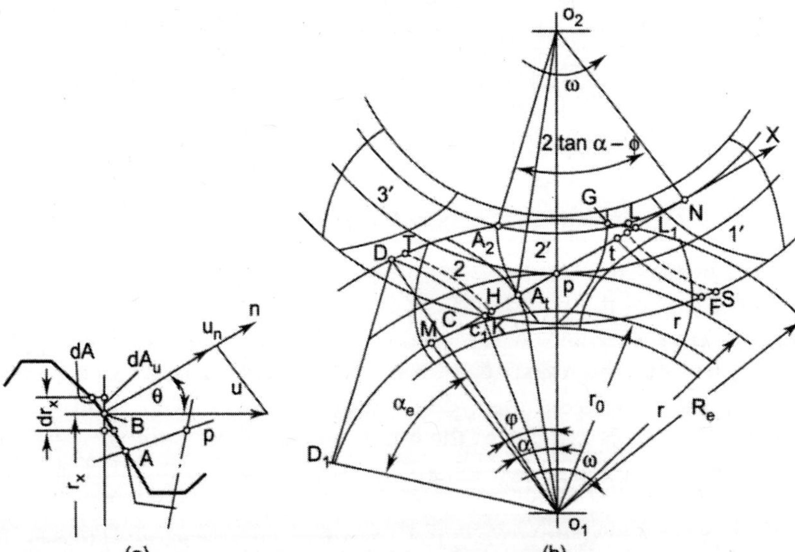

(a) (b)

Figure 5.5. (a) Calculation of instantaneous flow (b) Teeth engagement

Where R_e and r_a are the radius at the beginning point of the contact area and at the end point of the contact area of the teeth respectively. Taking the distance between the contact point A and the point P, AP = – x the radius r_A for driver and driven will be (figure 5.5*a*)

$$r_{AO1}^2 = r_0^2 + (MP - AP)^2 = r_0^2 + \left(\sqrt{r^2 - r_0^2} + x\right)^2$$

$$r_{AO2}^2 = r_0^2 + (NP + AP)^2 = r_0^2 + \left(\sqrt{r^2 - r_0^2} - x\right)^2 \qquad ...(5.6)$$

Where r and r_o are the beginning and base radius. Combining the above two equation (5.5 and 5.6) in order to change r_a^2

$$q_{1ins} = b\omega \frac{R_e^2 - r_{AO1}^2}{2} = \frac{b\omega}{2}\left(R_e^2 - r^2 - 2x\sqrt{r^2 - r_0^2} - x^2\right)$$

$$q_{2ins} = b\omega \frac{R_e^2 - r_{AO2}^2}{2} = \frac{b\omega}{2}\left(R_e^2 - r^2 - 2x\sqrt{r^2 - r_0^2} - x^2\right)$$

$$q_{ins} = q_{1ins} + q_{2ins} = b\omega\left(R_e^2 - r^2 - x^2\right) \qquad ...(5.7)$$

Referring to figure (5.5)

$$R_e^2 = \frac{r_0^2}{\cos^2 \alpha_e} = r_0^2(1 + \tan \alpha_e)^2 ; r^2 = \frac{r_0^2}{\cos^2 \alpha} = r_0^2(1 + \tan \alpha)^2$$

$$MP = MA + AP = \text{arc } MA_1 - x = r_0\phi - x = r_0 \tan \alpha$$

$$x = -r_0(\tan \alpha - \phi) \qquad ...(5.8)$$

Flow at any angle of rotation ϕ

$$q_{1ins} = \frac{b\omega r_0^2}{2}\left(\tan^2 \alpha_e - \phi^2\right)$$

$$q_{2ins} = \frac{b\omega r_0^2}{2}\left(\tan^2 \alpha_e - (2\tan \alpha - \phi)^2\right)$$

$$q_{ins} = b\omega r_0^2\left[\tan^2 \alpha_e - \tan^2 \alpha - (\tan \alpha - \phi)^2\right] \qquad ...(5.9)$$

The instantaneous flow as a function of angle ϕ is represented in the graphical form (line habcdef in figure 5.6).

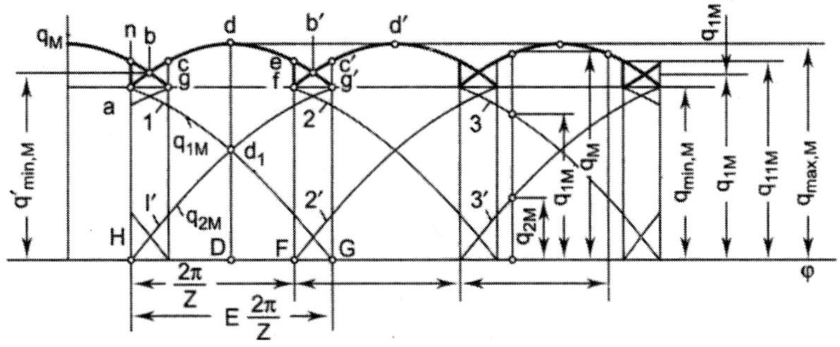

Figure 5.6. Graphical representation of flow in gear pump

At the time of contact when $x = PC = \frac{\varepsilon t_0}{2}$

From the equation 5.8, $\phi = \tan\alpha - \frac{\varepsilon t_0}{2r_0} = \tan\alpha - \frac{\varepsilon\pi}{Z}$, then

$$q_{1ins} = q_{1max} = aH = \frac{b\omega r_0^2}{2}\left(\tan^2\alpha_e - \left(\tan\alpha - \frac{\varepsilon\pi}{Z}\right)^2\right)$$

$q_{2ins} = q_{2min} = 0$, (point H in figure 5.6).

$$q_{ins} = q_{min\,ins} = aH = b\omega r_0^2\left(\tan^2\alpha_e - \tan^2\alpha - \left(\frac{\varepsilon\pi}{z}\right)^2\right) = b\omega\left(R_e^2 - r^2 - \left(\frac{\varepsilon t_0}{2}\right)^2\right) \quad ...(5.10)$$

ε is the amount of arc length S engaged with pitch t

The coefficient $\varepsilon = \frac{s}{t}$, indicates how many pairs of the teeth are in contact. From figure 5.5*b*

Since $\quad S = \frac{s_0}{\cos\alpha}$, $\varepsilon = \frac{s_0}{t\cos\alpha} = \frac{s_0}{t_0} = \frac{cL_1}{t_0}$ where $t_0 = \frac{2\pi r_0}{Z}$

When the contact point passes through the polus point P,

$x = 0$ and $\phi = \tan\alpha$

$$q_{1m} = q_{2m} = d_1 D$$

$$= \frac{b\omega r_0^2}{2}\left(\tan^2\alpha_e - \tan^2\alpha\right)$$

$$= \frac{b\omega}{2}\left(R_e^2 - r^2\right)$$

$$q_{ins} = q_{max\,ins} = 2q_{1ins} = dD = b\omega r_0^2\left(\tan^2\alpha_e - \tan^2\alpha\right) = b\omega\left(R_e^2 - r^2\right) \quad ...(5.11)$$

Flow fluctuation co-efficient

$$\delta_{qins} = \frac{2q_{max\,ins} - q_{min\,ins}}{\left(q_{max\,ins} + q_{min\,ins}\right)} = \left(\frac{\varepsilon\pi}{Z}\right)^2 \cdot \frac{1}{\left(R_e^2 - r^2 - \frac{1}{2}\left(\frac{\varepsilon\pi}{Z}\right)^2\right)} = \left(\frac{\varepsilon\pi}{Z}\right)^2 \frac{1}{\left(\tan^2\alpha_e - \tan^2\alpha - \frac{1}{2}\left(\frac{\varepsilon\pi}{Z}\right)^2\right)} \quad ...(5.12)$$

Graph in figure 5.7 gives, the variation of $\delta_{qins} = f(Z)$ (Curve No1) /2, 7,31/

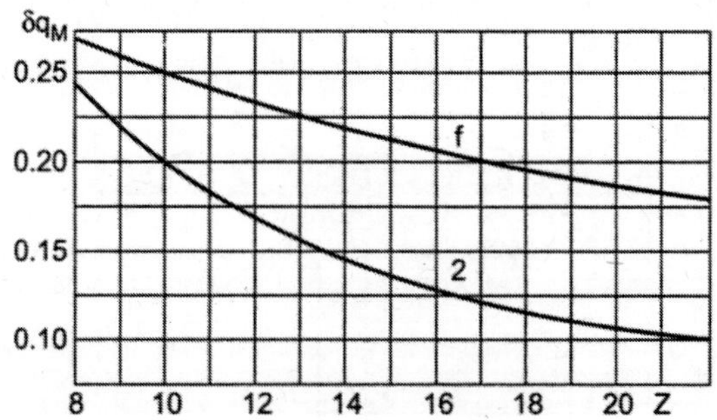

Figure 5.7. $\delta_{qins} = f(z)$

When the contact coefficient $\varepsilon = 1$, both teeth are in contact at more than one point, and fluid is entangled in between two teeth and gets compressed, (figure 5.8*a*) When $\varepsilon > 1$, contact takes place as in figure 5.8*b*. There will be a gap at point P by which the fluid occupies more space. This volume v_3 goes on changing and attains a minimum value Δv_3 when contacts are symmetrically located (figure 5.8*a*)

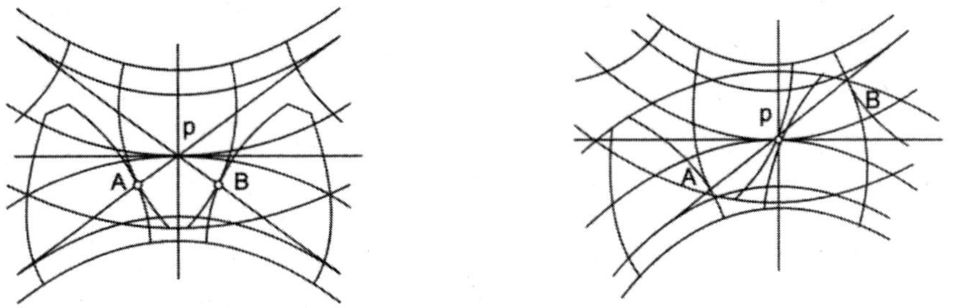

(a) Compression of the fluid between teeth, with contact points at the same tooth ε = 1 Two point contact on one tooth

(b) Compression of the fluid between teeth, with contact points at different teeth ε > 1 Three point contact on two teeth

Figure 5.8

The quantity of flow of this fluid in this space $q_{3ins} = \dfrac{dv_3}{dt}$ which is also equal to the difference between two instantaneous flow rate *i.e.*, $q_{3ins} = q_{IIins} - q_{Iins}$, where both contacts are located at point p at a distance, $x_I = -x$ and $x_{II} = t_o + x$

$$q_{3ins} = q_{IIins} - q_{Iins} = b\omega\left(R_e^2 - r^2 - (t_0 + x)^2\right) - b\omega\left(R_e^2 - r^2 - x^2\right) = -b\omega\left(2t_0\,x + t_0\right)$$

$$= b\omega r_0^2\,\frac{4\pi}{Z}\left(\tan\alpha - \frac{\pi}{Z} - \phi\right) \qquad ...(5.13)$$

Equation 5.13 is graphically represented in figure 5.7

q_{3max} is attained when $x = -\dfrac{\varepsilon t_0}{2}$

$$q_{3max\,ins} = b\omega t_0^2\,(\varepsilon - 1) = b\omega r_0^2\,\frac{4\pi^2}{Z^2}\,(\varepsilon - 1) \qquad ...(5.14)$$

When $x = -\dfrac{t_0}{2}$, $q_{3ins} = 0$. When $x = -\left(t_0 - \dfrac{\varepsilon t_0}{2}\right)$ and at this time $q_{3ins\,min} = -q_{3ins\,max}$

The volume Δv_3 is equal to volume ahba in figure 5.6

$$\Delta v_3 = \int q_{3ins}\,dt = \int_{-\frac{\varepsilon t_0}{2}}^{-\frac{t_0}{2}} \frac{bt_0}{r_0}\,(2x + t_0)\,dx = \frac{bt_0^3}{4r_0}\,(\varepsilon - 1)^2 = \frac{\pi b t_0^2}{2Z}\,(\varepsilon - 1)^2 \qquad ...(5.15)$$

Reduction of this space creates instantaneous high compression in fluid which induces an extra radial pulsating force along with the regular pressure force. During further rotation, increase in volume occurs from v_{3min} when $x = -\dfrac{t_0}{2}$ to v_{3max} when $x = -t_0\left(1 - \dfrac{\varepsilon}{2}\right)$. At this time, both the teeth leave the contact area.

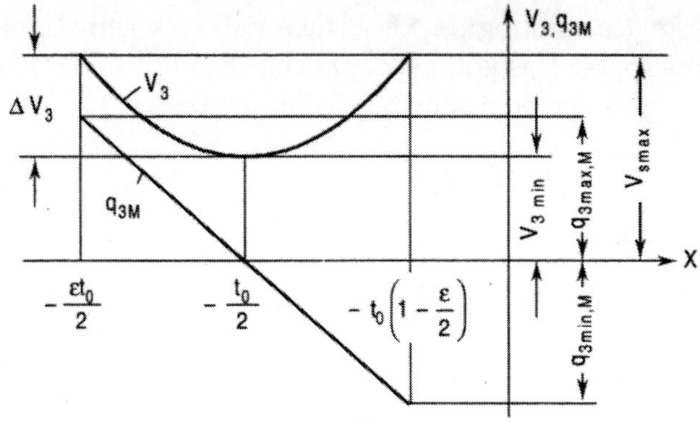

Figure 5.9. Graph $v_{3ins} = f(x)$ and $q_{max} = f(x)$

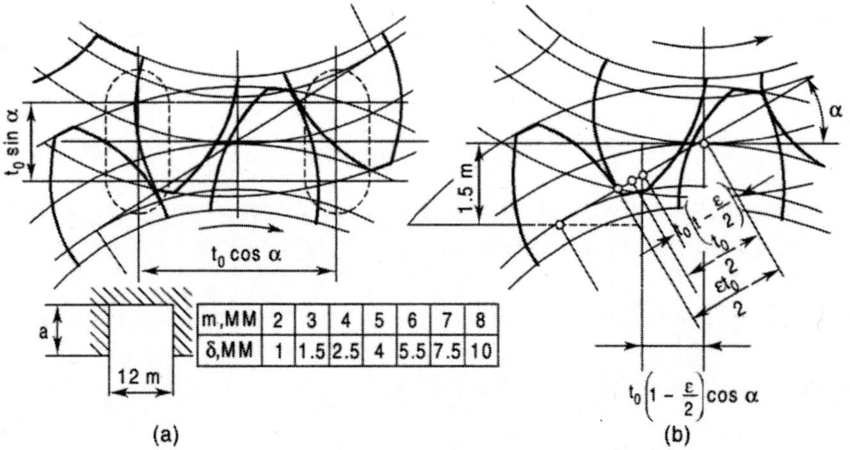

m,MM	2	3	4	5	6	7	8
δ,MM	1	1.5	2.5	4	5.5	7.5	10

Figure 5.10. Compression and Decompression by side ports

Pressure of fluid suddenly falls and air or vapour is released from the fluid depending upon the pressure. This vapour goes to the suction chamber. /11,16/ From the suction chamber, this vapour mixed liquid enters the space between two teeth of both the gears. The flow rate is reduced. This is removed by providing a number of radial holes at the root or base diameters and at the outer diameters of both the gears, connecting to a common chamber at the space near the shaft (figure 5.13) as well as by providing side holes at the pump side cover, figure 5.10 connecting the suction and delivery chambers to the compression and decompression area by which high compression is avoided (figure 5.13). However the flow rate is reduced. Although the radial holes method ensures efficient, due to high cost of manufacture and complicated manufacturing process, radial holes drilling in gears are rarely used, but side opening process is often used. The fluid v_3 at the time of reducing goes to delivery side due to the high pressure created and at the time of increasing goes to suction side due to the pressure reduction. There will be two holes symmetrically located (fig 5.10). These two holes will never be in contact with each other at any time of contact of both the teeth. The quantity v_3 is recirculated and the actual flow rate in the space between the gears is increased. The increase in volume Δv_3 reduces the fluctuation coefficient δ_{qins}.

Instantaneous flow during the contact period is given by the line hbcdeb′c′d′ (Fig. 5.6). The system of connecting the compression space only to delivery side holes ensures complete filling of the space between teeth when they come in contact with the suction chamber. But the volume Δv_3 is reduced. The line of instantaneous flow q_{ins} is indicated by hbgcde (Fig. 5.6).

$$\delta_{qins} = \frac{\left(\frac{t_0}{2}\right)^2}{\left(R_e^2 - r^2 - \frac{1}{2}\left(\frac{\varepsilon t_0}{Z}\right)^2\right)} = \frac{\pi^2}{Z^2\left(\tan^2\alpha_e - \tan^2\alpha - \frac{1}{2}\left(\frac{\pi}{Z}\right)^2\right)} \quad ...(5.16)$$

Fluid compression can be neglected in the compression space, provided the compression space and the corresponding volume are not reduced. In order to achieve this, the contact must start, when the volume is minimum.

Shifting of the starting point of contact can be obtained,

1. by reducing the outer diameter of the driven gear so that the outer circle at the time of release crosses the line of contact at a distance of $(t_o/2)$ from the point P.
2. By increasing the distance between the gear centers and
3. By rounding off the top portion of the teeth of the driven gears at the working side of the gear surface. However this is more difficult process than providing side channel as in figure 5.10, 5.12 and hence not adopted.

5.6 COMPRESSIBILITY EFFECT OF THE LIQUID BETWEEN TWO POINTS OF CONTACT

Theoretical analysis and experimental results indicate that the value of the coefficient ε for spur gear pumps is ε > 1.088 to 1.10. Failure to confine this value increases the wear of the teeth. Continuous, smooth contact between the teeth could not be achieved. Spur gears are always manufactured with ε > 1. /5,6/ At this condition, a portion of the liquid is compressed in the space between two contact points of the driver and the driven gears. Referring to figure 5.11(*a*), the space between two teeth 1 and 2 of the driven gear is in contact with delivery chamber (2), but not with the suction chamber (6) due to the contact of the driver with the driven, at the point **'*a*'**. For further rotation of gears, contact between the driver and the driven gears takes place at 3 points, as a result of which the liquid in the space between teeth and contact points **'*a*'** is under compression. (Figure 5.11(*b*)). In figure 5.11(*c*), it can be observed that the area of one compression space is reduced, where as the other compression space is opened to suction chamber. The magnitude of this pressure in this compressed space is instantaneously increased due to the rotation of the gears and also very high since the liquid is incompressible. This occurs when the axis of the tooth coincides with the line, joining the two gear centers (figure 5.11(*c*)). During further rotation, this pressure is reduced to minimum. Sometimes this pressure goes down below vapour pressure, as a result of which, a vacuum or decompression is created. Due to the decompression, liquid becomes vapour.

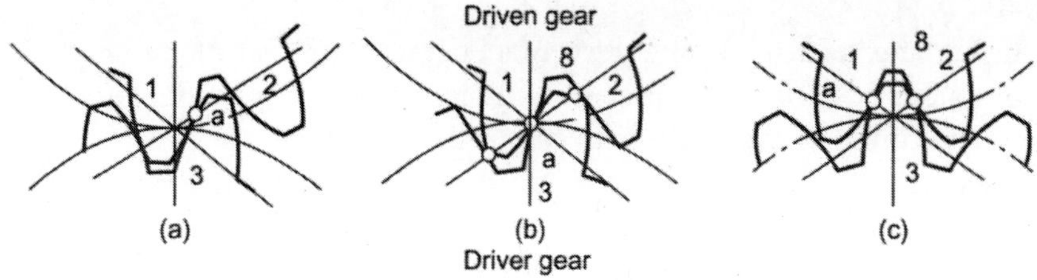

Figure 5.11. Liquid compression between teeth at ε = 1

This vapour, along with the liquid, goes to the suction space through the opening between the two gears 1 and 3, figure 5.11(*b*). This reduces actual quantity of flow. Further, this compressive force exerts a bending force on the shaft and between the outer surface of the gear, and the inner surface of the housing. Suction conditions also become poor.

All these increase the wear of the teeth profile, wear at the outer circumferential surface, both on the gear and on the housing, increase in leakage flow through the radial clearance between the gear and the housing, the volumetric efficiency and actual flow rate reduce. The mechanical losses are increased, due to the one sided impulsive load created on the bearings. The suction conditions become poor. Overall efficiency is reduced. Due to wear of the gear and the housing, the life of the gear pump is also reduced.

5.7 COMPRESSION AND DECOMPRESSION PROCESS IN GEAR PUMPS

The liquid, filled up in the space between driver gear tooth and driven gear tooth at the engaging point *i.e.*, at non-pumping region, undergoes (figure 5.8) instantaneous compression and decompression process during rotation. This process takes place only when the contact points of the teeth are at more than one point ($\varepsilon > 1$) (figure 5.8). Liquid is compressed during further rotation, if the contact points lie on the same teeth, whereas, the flow increases and pressure decreases if the contact points are at the successive teeth. Very high pressure is created, instantaneously, during compression. This pressure acts on the bearings, especially on the driver and on the teeth itself. The shaft bends, due to the fatigue stress, created by this force. The radial and axial clearances reduce at one end and increase at the other end. The wear on the gear and on the side cover increases. Volumetric and Mechanical efficiencies decrease. The liquid also gets heated.

When pressure reduces, vapour comes out from this hot liquid and occupies large space, which in turn reduces the flow rate.

5.8 METHODS TO REDUCE THE COMPRESSION AND DECOMPRESSION

To overcome this drawback, straight spur gears are used, for pumps with low flow rate. The contact ratio ε is slightly greater than $\varepsilon > 1$. The clearance between teeth and the housing lies between 0.1 and 0.4 mm. For pumps of higher flow rate, inclined gears or Herring bond gears are used, by which, the compression and decompression effect within the gears are reduced to a minimum.

In order to reduce the compression and decompression effect in pumps of low and medium flow rate, having spur gears, pressure releasing holes are provided, at the bottom between gears and/or at the outer surface of the gears. These holes are connected to a common space. During compression the liquid from the compressed space between gears flows to the common space, and during decompression the liquid flows from the common chamber to the decompressed space between gears. The pressure fluctuations are reduced considerably. In some designs, width of alternate tooth is reduced slightly to get cushioning effect. However, this method is adopted very rare. Volumetric efficiency reduces too much due to high rate of internal circulation. This method is not adopted in high-pressure pumps (above 50 to 70 ata), especially when liquid possesses a low viscosity.

In figure 5.12, I, II two positions of contact between teeth are illustrated under the condition $\varepsilon > 1$. Pockets **I** and **II** refer to the openings provided at both sides of the gears, *i.e.*, at the side covers. Due to the location of these ports, the liquid under compression and decompression conditions is connected either to the delivery chamber, through port **I** or to the suction chamber through port **II** respectively. Referring to figure 5.10(*a*), the compressed liquid in the space between the teeth during rotation of the gears, is connected to the delivery chamber, by which the compressed liquid flows into the delivery chamber there by compression, is avoided. In figure 5.10(*b*), the same compression space is connected to the suction chamber there by liquid from suction chamber flows into the decompression area. This ensures even pressure distribution, which reduces the wear of the tooth profile and other parts of gear pump. /2,16,22/

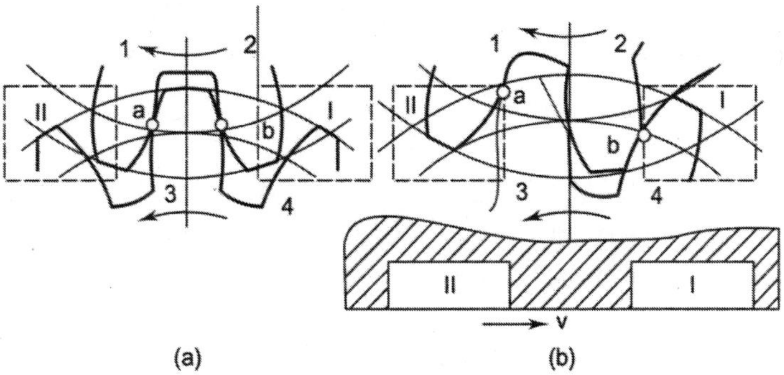

(a) (b)

Figure 5.12. Compression and decompression of liquid between teeth at $\varepsilon > 1$

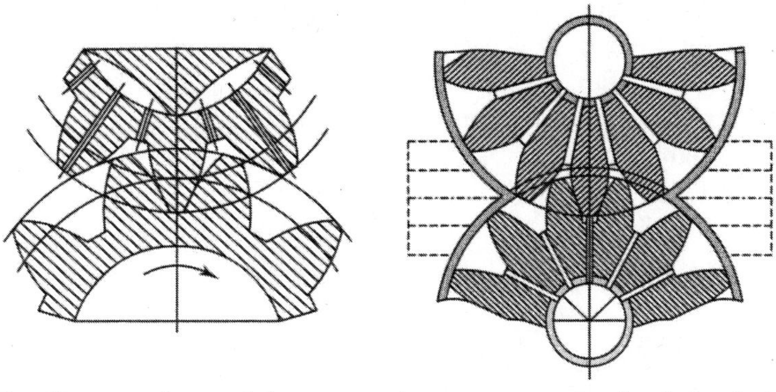

Figure 5.13. Compression and decompression pressure releasing holes between teeth

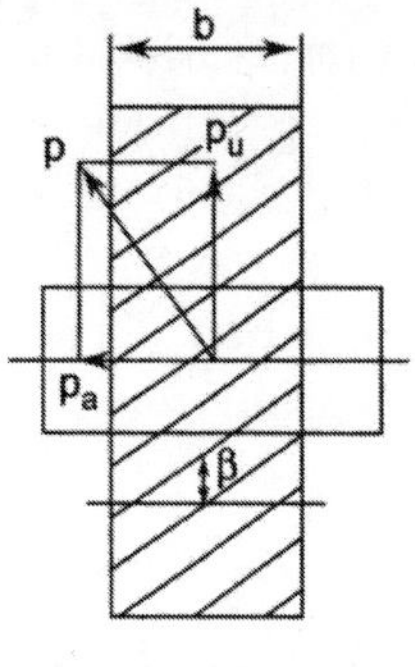

Figure 5.14. Gear with inclined teeth

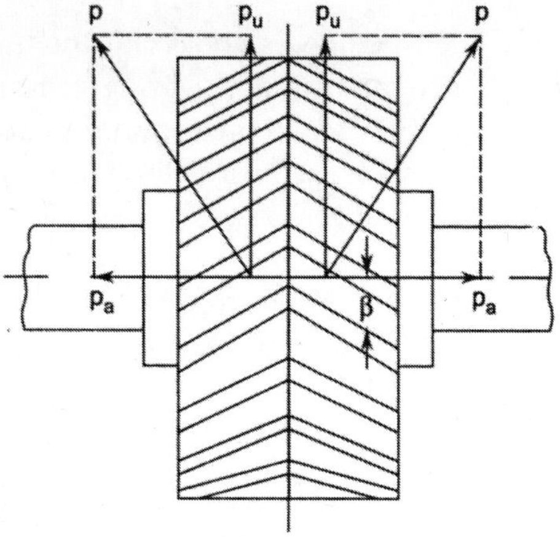

Figure 5.15. Gear with Herring Bond type teeth

Most radical method of reducing the compression and decompression is by having an angular contact **spur gear arrangement** as shown in figure 5.14. In this method while one end is at the beginning of compression, the other end will be in decompression. Due to continuity between these two ends, the liquid flows from high pressure to low pressure region

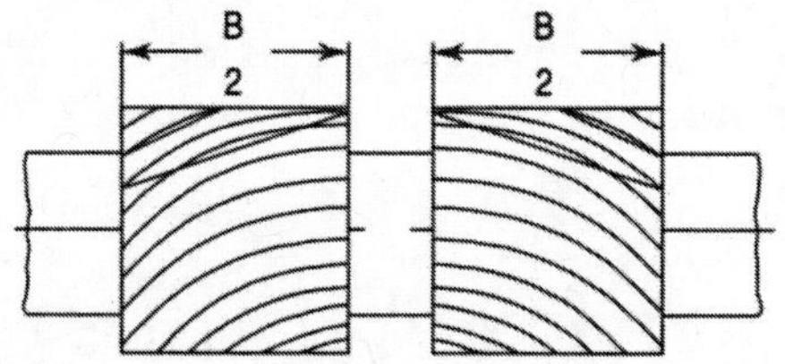

Figure 5.16. Two separate gears with mirror image Herring Bond type teeth

and the high pressure is released. The flow rate is uniform. The angle of inclination β of the teeth is selected between 20° and 30°. While selecting the value β, it is essential to see that there should not be a cross flow from delivery side to the suction side through the space in the gear. This method is found to be quite efficient and a stream lined uniform flow with higher volumetric efficiency is achieved. One disadvantage in this process is that axial force is developed (figure 5.15) which exerts a load on shaft and bearings.

In order to balance the axial force, a Herring bond type two-sided inclined tooth as indicated in figure 5.15 is used. Here the axial forces are balanced and at the same time, smooth uniform flow is maintained. In order to simplify the manufacture of Herring bond type gears, two separate gears, one the mirror image of the other, are assembled (figure 5.16). Straight spur gears are used for a pressure of 40 to 100 ata. Herring bond type gears are used for pressures of 20 to 40 ata. Inclined gears are used for 3 to 5 ata. More uniform flow prevails in inclined and in Herring bond type gears. In general the uniformity in flow rate is attained by proper selection of angle of contact β and number of teeth Z.

5.9 FLOW RATE

Flow rate $$Q_p = Zn \int q\,dt$$

Substituting $dt = \frac{dx}{\omega r_0}$ and $q_m = b\omega\left(R_e^2 - r^2 - x^2\right)$

Flow rate, without including the flow in between two gears at the time of contact

$$Q_{p1} = Znb \int_{\frac{\varepsilon t_0}{2}}^{t_0 = \frac{\varepsilon t_0}{2}} \left(R_e^2 - r^2 - x^2\right)\frac{dx}{r_0}$$

$$= 2\pi bn\left[R^2 - r^2\left(\frac{4}{3} - 2\varepsilon - \varepsilon^2\right)\frac{t_0}{4}\right]$$

$$= 2\pi bn\left[\tan^2\alpha_e - \tan^2\alpha - \left(\frac{4}{3} - 2\varepsilon - \varepsilon^2\right)\left(\frac{\pi}{Z}\right)^2\right] \qquad \text{...(5.17(a))}$$

Flow rate including the flow in the space between two gears at the time of contact

$$Q_{p2} = Znb \int_{\frac{t_0}{2}}^{\frac{t_0}{2}} \left(R_e^2 - r^2 - x^2\right)\frac{dx}{r_0}$$

$$= 2\pi bn\left[R^2 - r^2 - \frac{t_0^2}{12}\right]$$

$$= 2\pi bnr_0^2\left[\tan^2\alpha_e - \tan^2\alpha - \frac{\pi^2}{3Z^2}\right] \qquad \text{...(5.17(b))}$$

The extra quantity of flow available due to the induction of flow in the space between two gears at the time of contact

$$\Delta Q = Q_{p2} - Q_{p1} = \pi bn\frac{t_0^2}{2}(\varepsilon - 1)^2 = 2\pi bnr_0^2 \cdot \left(\frac{\pi}{Z}\right)^2(\varepsilon - 1)^2 \qquad \text{...(5.17(c))}$$

However an approximate formula for flow rate with considerable accuracy can be written as

$$Q_p = 4\pi rbn(R_e - r) \qquad \text{...(5.18)}$$

Actual quantity $\mathbf{Q}_a$ will be less than $\mathbf{Q}_{th}$, due to leakage quantity q through

(*a*) the radial clearance between gears 1 and casing 2,

(*b*) the axial clearance between gears and covers 3, as well as

(*c*) between teeth during contact period.

However maximum leakage flow takes place through axial clearance. Radial clearance is recommended as $\Delta r = 0.0015D_e$ (normal value is between **0.15 mm and 0.2 mm**). It is found that the fluid friction between the gear surface and the casing surface is very small. In order to have a dependable working of the pump, without quick wear of the pump parts, axial clearance Δ_T is selected larger than **0.15 mm** on both sides (normal value is between **0.1 and 0.15**) for normal spur gears. /7, 16/

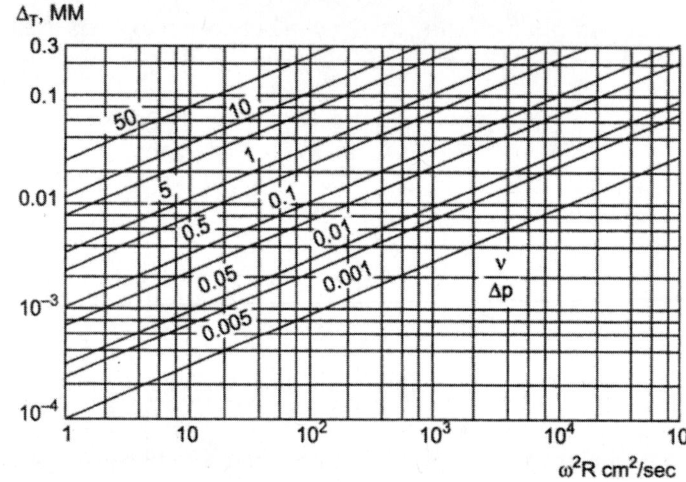

Figure 5.17. Monogram for the selection of side clearance Δ_T

If the teeth are normal and uncorrected $\mathbf{2r = mZ}$ and $\mathbf{2\ R_e = m\ (Z + 2)}$ where m is the module and **Z** is the number of teeth, the flow rate,

$$Q_{th} = 2p\, m^2 Zbn \qquad ...(5.19)$$

From the above equation it is found that, for the same radius of the gear R_e, higher module with less number of teeth will give higher flow rate. Based on this, the gear pumps are made with m = **4 to 6** and the number of teeth **Z = 8 to 14**. The nose of the gear wears out at a faster rate, if smaller number of teeth are selected. To avoid this, corrections are made on the teeth.

5.10 PROFILE CORRECTION

Profile correction is made by increasing the angle of contact α. Actual distances are increased from the calculated values by $\xi\, m$ (Figure 5.18)./3,5,6,21/

The profile correction and the rounding of the edge, reduce the value ε improve cavitation property, reduce compression effect and the friction loss, as well as increase the tooth strength. The correction coefficient ε_c is selected for the condition that, rounded corner for the teeth is not carried out and the thickness S_e is determined as $S_e = (0.2 \text{ to } 0.25)m$. Rounding of the edge is not carried out, if the distance between the centers of two identical gears with Z > 5 is equal to the centre distance between gears having number of teeth more than the designed value. Correction coefficient is selected as $\xi_B = \xi_c = 0.5$, $A = 2r = m\ (Z + 1) = m\ (Z + 2\xi_B)$.

Gears with $\xi_B = \xi_c = 0.5$ will be having the side clearance $C_\pi \approx (0.2 \text{ to } 0.1)m$, with **Z = 9 to 22.** In order to get any other value of clearance at $\xi = 0.5$, for example $\mathbf{C_s = 0.075\ m}$, the value of the coefficient ξ_c is determined from the equation for finding C_s (column 11 in the table 1) and then the value $\xi_c m$. The initial diameter d and **De** are determined for the $\xi_B = \xi_c = 0.32$ for Z = 33.

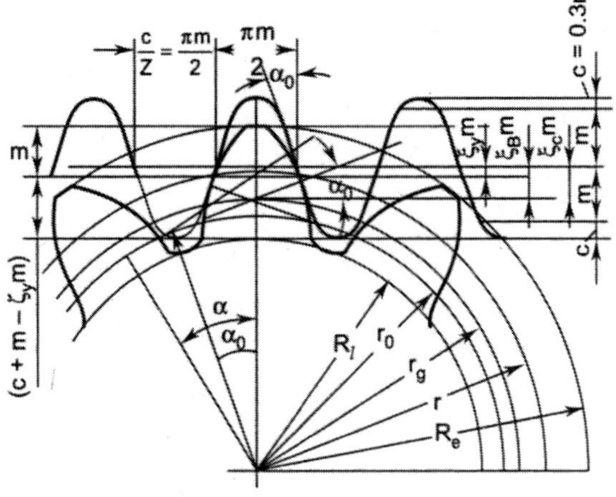

Figure 5.18. Scheme for the contact with corrected gears

Table 5.2 gives the calculation procedure of spur gears with correction.

Table 5.2. Calculations for spur type gear with profile correction.

1.	diameter of profile	d	$d = m(Z + 2\xi_B)$
2.	Angle of Contact	α	$\alpha = \arccos\left(\frac{Z}{Z + 2\xi_B}\cos\alpha_0\right) \alpha_0 = 20°$
3.	Outer diameter of the teeth	D_e	$D_e = m(Z + 2 + 2\xi_B)$
4.	base (root) diameter	D_i	$D_i = m(Z - 2.6 + 2\xi_B)$
5.	Pitch circle diameter	d_g	$d_g = mZ$
6.	Basic diameter	d_0	$d_0 = mZ . \cos\alpha_0$
7.	Pitch	t_0	$t_0 = \pi m . \cos\alpha_0$
8.	Co-efficient	ε	$\varepsilon = \frac{\sqrt{D_e^2 - d_0^2} - d\sin\alpha}{t_0}$
9.	Pressure angle at circumference	α_e	$\alpha_e = \arccos\left(\frac{d_0}{D_e}\cos\alpha_0\right)$
10.	Thickness	S_e	$S_e = D_e\left(\frac{\pi}{2Z} + \frac{2\varepsilon_c}{Z}\tan\alpha_0 + \text{inv}\,\alpha_0 - \text{inv}\,\alpha_e\right)$ $\text{inv}\,\alpha_0 = \tan\alpha_0 - \alpha_0,\ \text{inv}\,\alpha_e = \tan\alpha_e - \alpha_e$
11.	Side clearance between tooth profile	C_π	When $\xi_B = \xi_c C_\pi = 2Zm\cos\alpha_0\left[\text{inv}\,\alpha_0 - \text{inv}\,\alpha_e - \frac{2\xi_c}{Z}\tan\alpha_0\right]$

Correction improves the quality of the gear pump under the condition $\xi_B = 0.5$ ($\xi_c = \xi_B + \xi_y$). British manufacturers take $\xi_B = 0.02\,(30 - Z)$. European manufacturers adopt more number of teeth than minimum number of teeth calculated as per the equation $\xi_i = \xi_B = 0.33/20,14,29/$

For Z = 33, and $Z_{min} = \dfrac{2}{\sin^2 \alpha_0}$. This equation is applied for corrected profiles.

5.11 GEARS WITH INCLINED TEETH

Spur gears **with inclined teeth** and **Herring bond** gears are found to be smooth and noiseless in operation. In order to avoid the liquid, entering the suction chamber from delivery chamber through the space between the two gears at the contact point, the contact between the teeth must always exist continuously, This is possible, when

$$\varepsilon_s t_d > t_d + b \tan \beta_d$$

or

$$\tan \beta_d < \frac{2\pi rd}{bz} (\varepsilon_s - 1) \qquad ...(5.20)$$

where β_d is the angle of inclination of the screw length of the teeth along the axial direction on the cylinder, ε_s is the coefficient and $t_d = \dfrac{2\pi rd}{z}$ is the pitch on the cylinder

Instantaneous flow q_{ii} through one inclined gear is determined from the formula

$$q_{ii} = \int_a u\, da\,. \text{Since } u_n = u \cos \beta \cos (n, u)$$

$$q_{ii} = \int_a u_n\, da .= \int_a u \cos \beta \cos(n, u) da = \int_{au} u\, da_u = \int_{r_0}^{R_e} \int_0^b u\, dr_x\, dy$$

where $da_u = da \cos \beta \cos (n, u) = dr_x\, dy$ the projection of the elementary area da of the teeth along the plane passing through the point B and the centre of the gear (figure 5.20) On integration.

$$q_{iim} = \int_0^b \int_{r_0}^{R_e} r_x\, \omega dr_x\, dy = \int_0^b \omega \frac{(R_e^2 - r_B^2)}{2} dy \qquad ...(5.21)$$

The radii r_B at the point B on the teeth surface of the driver and the driven are equal (figure 5.20)

$$r_{BO_1}^2 = BO_1^2 = MO_1^2 + MB^2 = r_0^2 + (MP - AP + AB)^2$$

$$= r_0^2 + \left(\sqrt{r^2 - r_0^2} + x + \frac{r_0 y}{r_d} \tan \beta_d\right)^2 \qquad ...(5.22)$$

$$r_{BO_2}^2 = BO_2^2 = NO_2^2 + NB^2 = r_0^2 + (NP + AP + AB)^2$$

$$= r_0^2 + \left(\sqrt{r^2 - r_0^2} - x - \frac{r_0 y}{r_d} \tan \beta_d\right)^2 \qquad ...(5.23)$$

where AP = x, the distance between the contact point of both pitch circle diameters of both the gears p and the front edge of the contact point of both the gears.

$$AB = \frac{r_0 \, y}{r_d} \tan \beta_{d,}$$

The distance between two contact points along the line of contact, located at the front side of the gear and the section located at a distance of y.

Substituting the values for r_{B0}^2, and r_{B0}^2, in the equation 5.10,

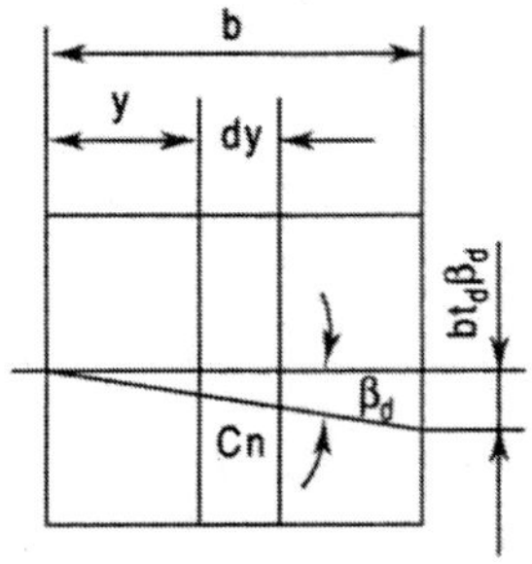

Figure 5.19. Calculation of angle of inclination β_d for the gears

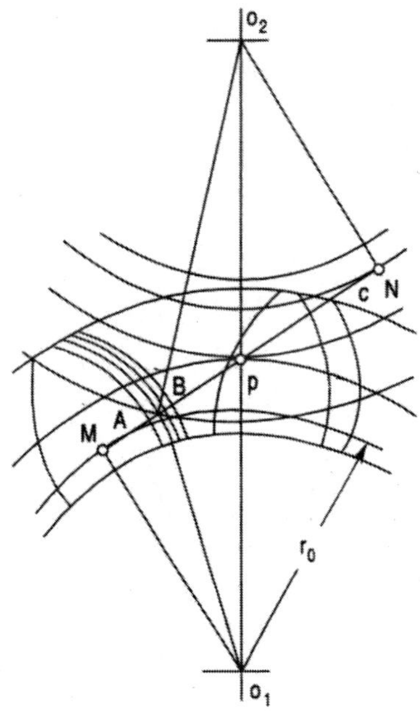

Figure 5.20. For the calculation of instantaneous flow through the inclined teeth

$$q_{i1,M} = \frac{b\omega}{2}\left(R_e^2 - r^2 - x^2 - \left(2\sqrt{\left(r^2 - r_0^2\right)} + \frac{br_0}{rd}\tan\beta_d\right)x - \sqrt{\left(r^2 - r_0^2\right)}\,\frac{br_o}{rd}\tan\beta_d - \frac{b^2 r_0^2}{3r_d^2}\tan^2\beta_d\right) \quad ...(5.24)$$

$$q_{i2,M} = \frac{b\omega}{2}\left(R_e^2 - r^2 - x^2 + \left(2\sqrt{\left(r^2 - r_o^2\right)} - \frac{br_0}{rd}\tan\beta_d\right)x + \sqrt{\left(r^2 - r_0^2\right)}\,\frac{br_o}{rd}\tan\beta_d - \frac{b^2 r_0^2}{3\, r_d^2}\tan^2\beta_d\right) \quad ...(5.25)$$

$$q_{i2,\,M} = q_{i1,\,M} + q_{i2,\,M} = b\omega\left(R_e^2 - r^2 - x^2 - \left(\frac{br_0}{rd}\tan\beta_d\right)x - \frac{b^2}{3}\frac{r_0^2}{r_d^2}\tan^2\beta_d\right) \qquad ...(5.26)$$

Since $R_e^2 - r^2 = r_0^2 - \left(\tan^2\alpha_e - \tan^2\alpha\right)$ and $x = -\,r_0\,(\tan\alpha - \varphi)$

$$q_{i2,\,M} = b\omega r_0^2\left(\tan^2\alpha_e - \tan^2\alpha - (\tan\alpha - \varphi)^2 + \frac{b}{r_d}\tan\beta_d(\tan\alpha - \varphi) - \frac{b^2}{3}\frac{\tan^2\beta_d}{r_d^2}\right) \qquad ...(5.27)$$

Instantaneous flow $q_{i,\,M}$ will be maximum when $x = -\frac{b}{2}\frac{r_0}{r_d}\tan\beta_d$. At this point the central point of the contact line of the teeth lies on the point P.

$$q_{i\max M} = b\omega\left(R_e^2 - r^2 - \frac{b^2}{12}\frac{r_0^2}{r_d^2}\tan^2\beta_d\right) \qquad ...(5.28)$$

and will be minimum, when $x = \frac{S_s}{2}t_0 + \frac{b}{2}\frac{r_o}{r_d}\tan\beta_d$

$$q_{i\min,\,M} = \frac{b\omega}{2}\left(R_e^2 - r^2 - \frac{s_s^2\,t_0^2}{4} - \frac{b^2}{12}\frac{r_0^2}{r_d^2}\tan^2\beta_d\right) \qquad ...(5.29)$$

Fluctuation of the instantaneous flow $A_i = q_{i\max,\,M} - q_{i\min,\,M}$

For inclined gears $A_i = \left(\frac{\varepsilon_s t_0}{2}\right)^2\omega b$ and for spur gears $A_s = \left(\frac{\varepsilon t_0}{2}\right)^2\omega b$

The coefficient of un equal flow

$$\delta_{qiM} = \frac{\left(\frac{\varepsilon_s t_0}{2}\right)^2}{\left(R_e^2 - r_0^2 - \frac{1}{2}\left(\frac{\varepsilon_s t_0}{2}\right)^2 - \frac{b^2}{12}\left(\frac{r_0}{r_d}\right)\tan^2\beta_d\right)} \qquad ...(5.30)$$

The flow Q_i will be

$$Q_i = Zn\int q_{i,\,M}\,dt = \frac{Znb}{r_0}\int_{\frac{\varepsilon_0 t_0}{2}}^{t_0 = \frac{\varepsilon_s t_0}{2}}\left(R_e^2 - r^2 - x^2 - \frac{br_0}{rd}\tan\beta_d\,.\,x - \frac{b^2}{3}\left(\frac{r_0}{r_d}\right)^2\tan^2\beta_d\right)dx \qquad ...(5.31)$$

$$Q_i = 2\pi bn\left[R_e^2 - r^2 - \frac{b^2}{3}\left(\frac{r_0}{r_d}\right)^2\tan^2\beta_d - \frac{br_0}{r_d}\tan\beta_d\,\frac{t_0(1-\varepsilon_s)}{2} - \left(4 - 6\varepsilon_s + 3\varepsilon_s^2\right)\frac{t_0^2}{12}\right] \qquad ...(5.32)$$

Since $\frac{br_0}{r_d}\tan\beta_d = t_0\,(\varepsilon_s - 1)$

$$Q_i = 2\pi bn\left[R_e^2 - r^2 - \left(\varepsilon_s^2 - 2\varepsilon_s - 2\right)\frac{t_0^2}{12}\right] \qquad ...(5.33)$$

5.12 FLOW RATE CALCULATION IN A GEAR PUMP

For each revolution of the gear, a quantity equal to the volume of the space between two successive gear teeth is transferred from suction space (6) to delivery space (3). If b is the width of the gear parallel to the axis, *i.e.*, perpendicular to the figure 5.21, A is the area of the space between two successive teeth, Z is the number of spaces, which are equal to Z number of teeth in each gear and n is the speed in rpm, the total quantity for two gears, is

$$Q = \frac{2Zb\,An}{60}\,\frac{m^3}{\text{sec}} \qquad ...(5.34)$$

Due to curved nature of the gear teeth, the correct value of the area between two successive teeth can be determined approximately.

The area A can be expressed as $A \approx \frac{t}{2} \times h$, where $\boldsymbol{h}$ is the height of tooth, and $\boldsymbol{t}$ is the pitch between the teeth, $t = \frac{\pi d_p}{Z}$, where $\boldsymbol{d_p} = 2r_p$ is the pitch circle diameter. The height h is 2 times module m *i.e.*, $h = 2m = 2\frac{d_p}{Z}$. Expressing the height 'h' in terms of module $\boldsymbol{m}$ and pitch circle diameter of the gear $\boldsymbol{d_p}$, the area of the space between the two gears is

$$A = \frac{t}{2} \times h = \frac{\pi d_p}{Z}\,\frac{2d_p}{Z} = \frac{2\pi d_p^2}{Z^2}$$

and

$$Q = \frac{2Zb\,\pi d_p^2 n}{2Z^2 \,.\, 60} = \frac{\pi d_p^2 bn}{60Z}\,\frac{m^3}{\text{sec}} \qquad ...(5.35)$$

Based on the work done by Prof Toma / 20 /, Prof E.M.Udhin developed an expression for the flow rate in gear pump. The basic assumptions are

1. At a time, only one tooth of driver is in contact with one tooth of the driven.
2. The contour of the gear teeth is an involute. In such contours, minimum energy loss occurs. The energy supplied by the gear teeth can be taken as equal to the energy absorbed by the fluid. Referring to (figure 5.21), an unequal hydraulic pressure exists along the path $\boldsymbol{r_2 - x}$ on the line O_1E and along the path $\boldsymbol{r_2 - y}$ on the line O_2M.

The net pressure acting on the fluid will be $p = p_d - p_s$ where $\boldsymbol{p}$ is the net or total pressure, $\boldsymbol{p_s}$ is the suction pressure and $\boldsymbol{p_d}$ is the delivery pressure.

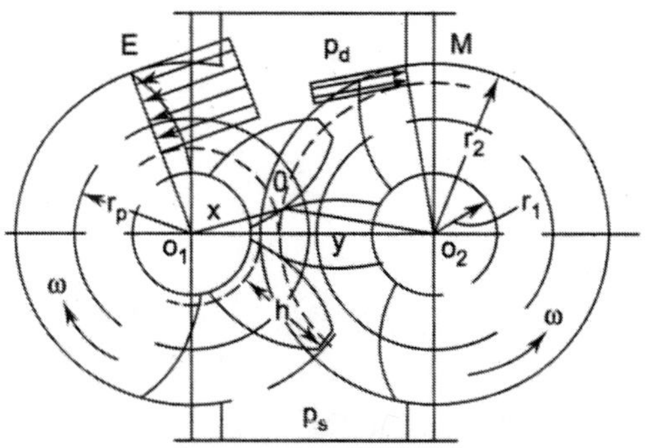

Figure 5.21. Flow calculations for a gear pump

Angular moment at the driver with respect to the axis O_1,

$$M_1 = p(r_2 - x)\, b\, \frac{r_2 + x}{2} = \frac{p.b}{2}\left(r_2^2 - x^2\right) \qquad ...(5.36)$$

Angular moment at the driven with respect to the axis O_2

$$M_2 = p(r_2 - y)\, b\, \frac{r_2 + y}{2} = \frac{p.b}{2}\left(r_2^2 - y^2\right) \qquad ...(5.37)$$

Total moment will be equal to the moment supplied by the prime mover and is determined at the gear end of the shaft.

$$M = M_1 + M_2 = \frac{p.b}{2}\left(2r_2^2\, x^2 - y^2\right) \qquad ...(5.38)$$

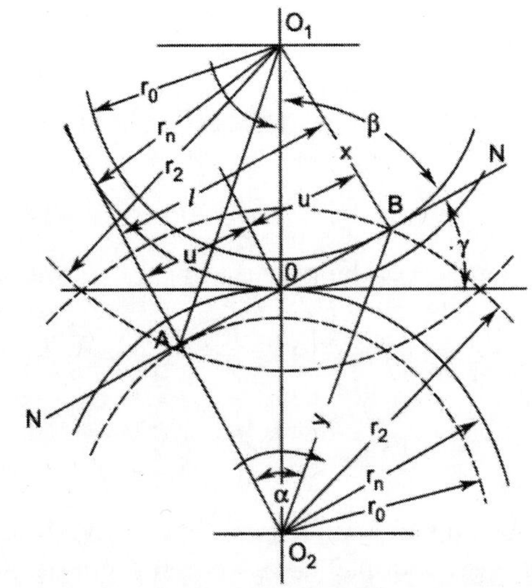

Line NN - Line of contact, r_2 – outer radius of the gear tooth, r_p – pitch circle radius, r_o – base circle radius, γ – angle of contact

Figure 5.22. Kinematic Scheme for gear mesh

The kinematic scheme of the contact between the teeth of the driver and that of the driven is given in the figure 5.38. (3, 5, 6) Considering the parallelogram O1A2B

$$x^2 + y^2 = 2\left(r_p^2 + u^2\right) \qquad ...(5.39)$$

Substituting this value in the equation 5.38 and rearranging

Total Moment, $$M = pb\left(r_2^2 - r_p^2 - u^2\right) \qquad ...(5.40)$$

But Moment $$M = \frac{N}{\omega} = \frac{Q_{th} \cdot p}{2\pi n} \qquad ...(5.41)$$

Combining both equations and simplifying

$$Q_{th} = \frac{M \,.\, 2\pi n}{p} = 2\pi b n\left(r_2^2 - r_p^2 - u^2\right) \qquad ...(5.42(a))$$

The flow rate, $\mathbf{Q_{th}}$, is the function of r_2, the outer radius of the gear, r_p the pitch circle radius of the gear and l the contact length, which changes with respect to the angle α.

Correspondingly the value u and the contact ratio ε also change. The pattern of the change of flow rate with respect to the angle α, can be studied first by taking ε = 1. This means that only one contact between the teeth is taking place at a time. At this time full teeth are in contact and the fluid is completely removed from the space between the teeth. At this condition $l = 0$, $u = 0$, and $\varepsilon = 1$ and the flow rate is maximum. The flow rate is minimum, at the beginning and at the end of the contact period. At this condition $u = l/2$, and $\varepsilon = 1$. Substituting these values, in equation 5.42(a)

$$Q_{th\,max} = 2\pi bn\left(r_2^2 - r_p^2\right) \quad \text{for } u = 0 \qquad \text{...(5.42(b))}$$

and

$$Q_{th\,min} = 2\pi bn\left(r_2^2 - r_p^2 - \left(\frac{l}{2}\right)^2\right) \quad \text{for } u = l/2, \qquad \text{...(5.43(a))}$$

The variation of $Q = f(\alpha)$ is shown in figure 5.23.

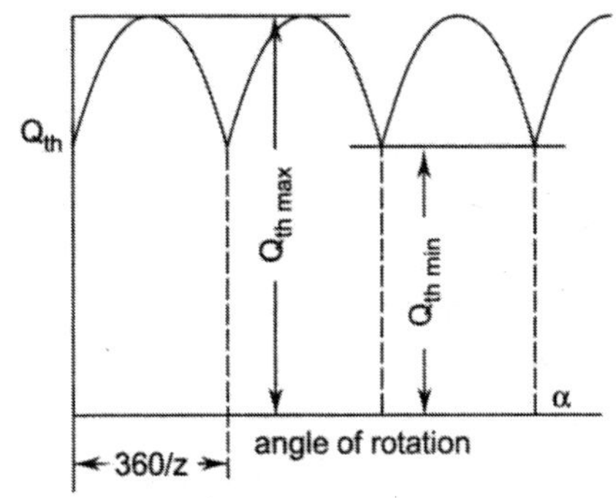

Figure 5.23. Flow in spur gear pump when ε = 1

The flow fluctuates and the % rate of fluctuation δ is given as

$$\delta = (Q_{th\,max} - Q_{th\,min})/Q_{th\,max}.$$

Since $r_2 = r_p + m = (mZ/2) + m,\ l = t\sin\beta = \pi m \sin\beta$, when $\varepsilon = 1$,

$$\delta = (\pi^2 \sin^2\beta)/4(Z+1)$$

Gear pumps, with more number of teeth will have lesser flow fluctuations, but the flow rate will be reduced. At the same time the liquid is compressed between teeth. Fluctuation of the flow can be reduced, by having more number of teeth with smaller width b, instead a few teeth with larger width b, and at the same time slightly turning the position teeth in one gear relative to the other by a small angle. The teeth profile becomes spiral or involute, instead of spur gear type.

Equations 5.43 (a) & (b) agree with the practical results when $u = 0$. When $u < 1$, the following formula given by Mustaris and Toma /31/can be used.

$$Q_{th} = 2\pi bn\left(r_2^2 - r_p^2 - k\left(\frac{t^2}{12}\right)\right) \qquad \text{...(5.43(b))}$$

Where t is the pitch of the gear teeth, $k = 4 - 6\varepsilon + 3\varepsilon^2 \approx 12$ for Z = 8 to 14, which is the most commonly used value.

5.13 SUCTION CHARACTERISTICS OF GEAR PUMP

Economy, dependence in operation, less weight and good suction conditions are the basic characteristics of a gear pump. Allowable speed for a gear pump is always selected based on the suction conditions, especially when the pumping liquid has considerable viscosity.

Applying Bernoulli's equation between the suction chamber and the fluid level in the suction tank (figure 5.24)

$$\frac{p_{at}}{\gamma} = \frac{p_1}{\gamma} + \frac{C_1^2}{2} + Z_1 + \sum h_{fs} + \Delta h \quad \text{...(5.44)}$$

where p_{at} is the atmospheric pressure or the pressure of the liquid in the suction tank p_1, C_1, Z_1 are the pressure, velocity and height of the pump above the liquid level in the suction tank, respectively,

$\Sigma\, h_{fs}$ is the sum of major and minor losses in the suction line, and

Δh is the loss in the suction chamber

If the pressure p_1 is equal to or less than vapour pressure p_{vp} of the liquid at the pumping temperature, vapour and gases comes out from the liquid. Cavitation starts, subsequent metal erosion takes place. Pump runs with noise and vibration and pumping is no more possible. Since the velocity C_1 at the suction chamber is zero,

$$\frac{p_{at}}{\gamma} = \frac{p_{vp}}{\gamma} + Z_1 + \Sigma h_{fs} + \Delta h$$

Maximum vacuum $h_{vac\,max}$ can be written as

$$h_{vac\,max} = \frac{(p_{at} - p_{vp})}{\gamma} = (Z_1 + \Sigma h_{fs}) + \Delta h \quad \text{...(5.45)}$$

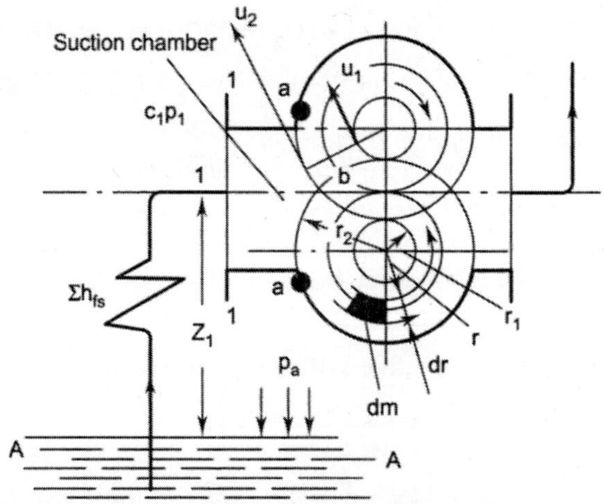

Figure 5.24. Determination of suction conditions in gear pump

Inside the chamber at the space between two teeth, the liquid should move at the velocity equal to the gear velocity. For the safe operation the pressure at the suction chamber must be above the vapour pressure, $p'_1 > p_{vp}$. Hence

$$\frac{p_1^1}{\gamma}+\frac{u_2^2}{2g}=\frac{p_1}{\gamma}+\frac{C_1^2}{2g}+\Delta h$$

Correspondingly $h_{vac\ allow}=\dfrac{(p_{at}-p_1')}{\gamma}=(Z_1+\Sigma h_{fs})+\Delta h+\dfrac{u_2^2}{2g}$ ...(5.46)

For calculation purposes $p_1'=p_1$, since the pressure p_1' cannot be measured

$$h_{vac\ allow}=\frac{(p_{at}-p_1)}{\gamma}\approx(Z_1+\Sigma h_{fs})+\Delta h+\frac{u_2^2}{2g} \qquad ...(5.47)$$

Since $u_2=\dfrac{\pi d_z n}{60}$, the allowable speed for cavitation free operation

$$n\le A\frac{\sqrt[4]{\left(\dfrac{(p_{at}-p_1)}{\gamma}-Z_1-\Sigma h_{fs}-\Delta h\right)^3}}{\sqrt{Q_{th}}}\ \text{rev/min} \qquad ...(5.48)$$

Where A is a coefficient ≈ 2500 to 5500, the pressure, $p_1>p_{vp}$, and Q_{th} is the theoretical discharge in m^3/hr

High speeds can be selected for low discharge pumps in order to reduce the weight. But high speed is not recommended for high discharge pumps, due to cavitation problem, normal speeds for high discharge units will be n = 300 to 1000 rpm, and for low discharge units n = 2000 to 4000 rpm.

The dynamic depression Δh depends upon

(*a*) The hydraulic losses in the gear surfaces exposed to the suction space, change in the direction of flow and the shock losses due to the striking of the fluid on the gear surface.

(*b*) The hydraulic losses due to accelerated flow from the velocity C_1, in the suction chamber to the space between gear teeth.

(*c*) The centrifugal force, dp acting on the fluid and moving the fluid from the base radius, r_1 to the outer radius, r_2. Referring to the figure 5.24, the elementary centrifugal force, acting on the elementary area dr located at a radius of r will be

$$dp=\frac{\gamma}{g}\omega^2 r$$

For the entire area, by integration between the radii r_1 to r_2

$$\int_{p_{r2}}^{p_{r2}}dp=\frac{\gamma}{g}\omega^2\int_{r_2}^{r_2}r\,dr$$

from which

$$p_{r2}-p_{r1}=\frac{\gamma}{g}\frac{\left(u_2^2-u_1^2\right)}{2} \qquad ...(5.49)$$

The loss is equal to

$$h_{fc} = \frac{(p_{r2} - p_{r1})}{\gamma} = \frac{(u_2^2 - u_1^2)}{2g} \qquad \text{...(5.50)}$$

(*d*) The loss due to the inertia force, $h_{f\,iner}$, which prevails only on the spur gears and not on inclined or herring bond type gears, which is inversely proportional to the number of teeth Z.

However, the dynamic depression Δh is always calculated as equal to h_{fc}. All other values are neglected. Substituting this value in the equation 5.50, the speed is calculated.

Cavitation tests and its graphical representation, on gear pumps, are, as similar to other positive displacement pumps. The allowable suction height H_{sv} is normally selected as H_{sv} allowable = 3 to 7.5m

Increased axial clearance between gear and the side casing considerably reduces the cavitational characteristics. This is due to the movement of gas present in the fluid from the delivery to suction passing through this radial clearance and again from suction to delivery through the space between two teeth value, which reduces the flow rate. When the pressure reaches critical value, flow even becomes zero. This occurs especially when the pumping liquid is at a high temperature.

Based on the experiments conducted by many authors,/7,11,16,31/ the speed is calculated by the following formula.

$$n \le \frac{C}{\sqrt{Q_{th}}} \qquad \text{...(5.51)}$$

where C = 4000 to 10,000. Lower value is selected, when hydraulic losses are low, for the liquids of low vapour pressure and for the velocity u_2 = 2 to 5m/sec, and higher value is for higher vapour pressure, for higher hydraulic losses and the velocity u_2 = 5 to 8*m*/sec.

The suction chamber and the suction tube must also be without any gas presence. Otherwise, fluctuation of flow and subsequent impulsive force will act on bearings. Cavitation in pumps running at high speed can be avoided by running at lower speed and with reduced suction height or even with positive suction head.

5.14 LOSSES AND EFFICIENCY OF GEAR PUMPS

Mechanical losses in bearings, volumetric losses due to the leakage through the clearance and hydraulic losses due to friction are the main losses in gear pumps. Hydraulic losses are minimum in gear pumps since the flow velocity is minimum and the passage length, *i.e.*, the length of the surface of gear teeth and the casing inner surface is also very small. Hydraulic efficiency in gear pump lies between 98% and 99%.

Mechanical losses are also very small in magnitude, *i.e.*, the losses in bush or sleeve bearings, stuffing box. Volumetric losses occupy a major percentage. The volumetric efficiency $\eta_v \approx$ 58% to 96% for gear pumps.

5.14.1 Volumetric Losses (η_{fv}) and Volumetric Efficiency (η_v)

The axial clearance between gear and side cover plate, the radial clearance between the housing and the gear, high delivery pressure, high viscosities of oil are the major causes for high volumetric loss. Speed of rotation and flow rate influence the volumetric losses to a smaller extent.

Axial clearance between rotor and housing cover should be 0.15 to 0.2 mm while radial clearance between the housing inner diameter and gear outer diameter should be 0.225 to 0.25 mm. This clearance should be maintained constant independent of the size, flow rate and speed of the pump. The rate of increase in volumetric losses will be more especially when the radial clearance changes from 0.25 mm and above. If the axial clearance is increased further, the flow rate reduces to a larger extent. Sometimes the pump may not run. The leakage flow depends upon the clearance, viscosity of the fluid and the pressure difference between the suction and the delivery.

For medium and high discharge and high head pumps, (d_2 > 200mm) the clearance volume should be kept minimum. Metal erosion due the presence of the solids in suspension, metal erosion due to the jet action of the flow at the clearance, high pressure of liquid in the space between two gears, at the non-working area and subsequent compression and decompression of fluid, which creates a bending moment, as well as fatigue stress on the shaft, by which the radial and axial clearances reduce at one end and increase at the other end, which are the causes for the increase in clearance and corresponding increase in volumetric losses.

In high pressure gear pumps, due to high delivery pressure, the rate of flow through the clearance increases, which reduces actual flow rate and increases volumetric loss. Referring to figure 5.25, the line EM is for the minimum clearance and line EN is for the higher radial clearance (0.3 to 0.4 mm). Experiments show that axial clearance increase influences to a larger extent on volumetric loss than radial clearance increase. In addition, it is found, that the leakage flow rate is directly proportional to the delivery pressure./31/.

Increased viscosity of liquid reduces the flow rate through the clearance. At the same time, the space between the gear teeth is not filled completely due to slow movement of the pumping liquid, especially at high speeds and at high viscosity of the liquid. This reduces the actual flow rate, and volumetric efficiency. In gear pumps, (2/3) of the volumetric loss occurs due to incomplete filling of the space between gears, and 1/3 is due to leakage through clearance.

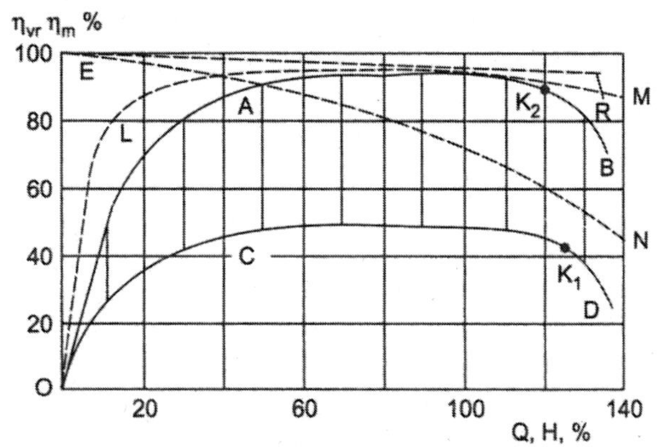

Figure 5.25. Variation of η_v and η_m = f(Q, H) expressed as % of optimum values

Apart from that, the wear of the gear tooth will be at a faster rate. The flow will not be uniform at the delivery. Delivery pressure also reduces. Cavitation may also occur at suction especially when the flow velocity exceeds 6 to 8 mps. Normally 1/8 part of the gear wheel is exposed to suction chamber and 1/8 part is exposed to delivery chamber for safe and efficient operation of pump. Absolute pressure of the fluid at suction chamber must be more than the centrifugal force, in order that the space is completely filled with liquid. Centrifugal force will be more, if viscosity is high.

Although increase in speed increases the volumetric efficiency, the space between two gears is not properly filled with liquid especially when viscosity is high, which reduces the volumetric efficiency.

Increase in flow rate increases the volumetric efficiency. The leakage flow through the clearance does not proportionately increase when flow rate is increased because of fixed value of clearance. The leakage flow is independent of the flow rate. Referring to figure 5.25 lines OAB and OCD correspond to volumetric efficiency. Line OAB represents $\eta_v = f(Q)$ for moderate flow rate (1 to 10 m^3/hr) and high head pumps (H = 150 to 400 MWC). From the graph, it can be seen that for the same increase in the flow rate, the volumetric efficiency drop is larger in one and less in the other.

The space between two gears is not filled completely when suction conditions are severe, which depends upon the liquid temperature and the speed of rotation.

Since volumetric efficiency is influenced by many parameters, different authors specify different equations. The most commonly adopted formula, which gives result with less percentage of error is,

$$\eta_v = 0.9\left(100 - \frac{7p}{\sqrt[3]{^0Eu}}\right)\% \qquad ...(5.52)$$

Also

$$\eta_v = A\left(100 - \frac{Bp}{R.u}\right)\% \qquad ...(5.53)$$

Where the experimental coefficients **A** = 0.9 to 1.0, **B** = 200 to 210, ***p*** is the net pressure between delivery and suction pipes (kg/cm^2) **°E** is the viscosity is angles **(°E) R** is the Redwoods in seconds, u is the contact length of two teeth at non-working area.

$$\eta_v = A\left(100 - \frac{108.3p}{R.u}\right)\% \qquad ...(5.54)$$

5.14.2 Mechanical Efficiency

Mechanical losses attribute to the losses in bearings, stuffing box, losses due to contact of the gear at the side as well as at the periphery of the casing. Mechanical efficiency ranges from 0.85 to 0.95 depending upon the flow rate of the pump. The line OLR in figure 5.25 indicates the change of mechanical efficiency as a function of % of flow rate to the normal flow rate (Q/Q_{opt}) at the upper limit of η_m = 95%. The same pattern of change takes place even when η_m = 85%. However, at high-pressure pumps the contact between casing and gear at different places as well as between two teeth is so high, the material of the gear and housing comes out as powder, which increases the mechanical losses, and also volumetric losses too much.

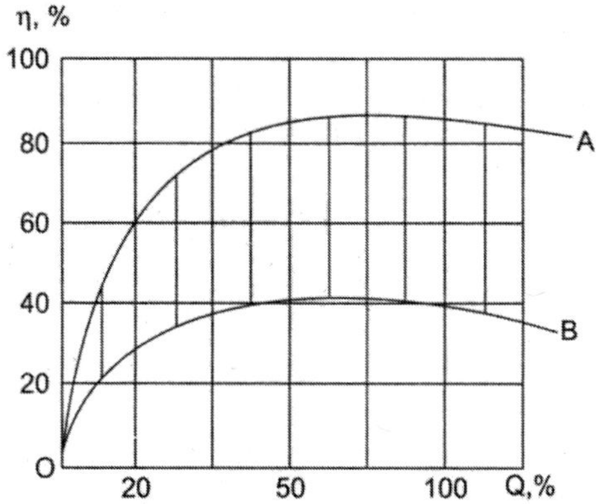

Figure 5.26. Overall efficiency of gear pump as a function of (Q/Q_{opt}) %

5.14.3 Overall Efficiency

Overall efficiency is the product of these three efficiencies. Overall efficiency of gear pumps ranges from 50% to 85% for stationary or land operations and 40% to 80% for the transport in mobile operation, such as automobile, ships, etc.

Figure 5.26 gives the variation of overall efficiency, from minimum (η = 40%) to maximum (η = 85%) with respect to the %flow rate (Q/Q_{opt}).

5.15 PERFORMANCE OF GEAR PUMP

Performance of a gear pump is represented in different forms, such as $Q = f(H)$, $N = f(H)$, $N = f(Q)$ and $N = f(n)$.

From the flow rate equation, it is seen that flow rate is proportional to the speed. In (Fig. 5.27), the performance graph $Q = f(n)$ is given. Line OA represents theoritical curve and line OB represents the actual curve. The difference between OB, ED, MN and OA is the volumetric loss, but the total head of operation is high for MN, low for ED and lower for OB, *i.e.*, $H_{MN} > H_{ED} > > H_{OB}$.

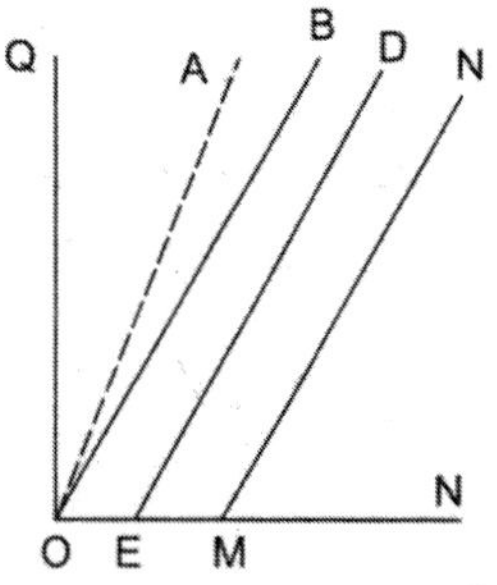

Figure 5.27. Performance of gear pump Q_{th} and $Q_a = f(n)$

5.16 FORCES ACTING ON GEAR PUMP

Forces acting on the gear pump and on its parts are the net difference in pressure between delivery and suction.

$$p = (p_d - p_s)$$

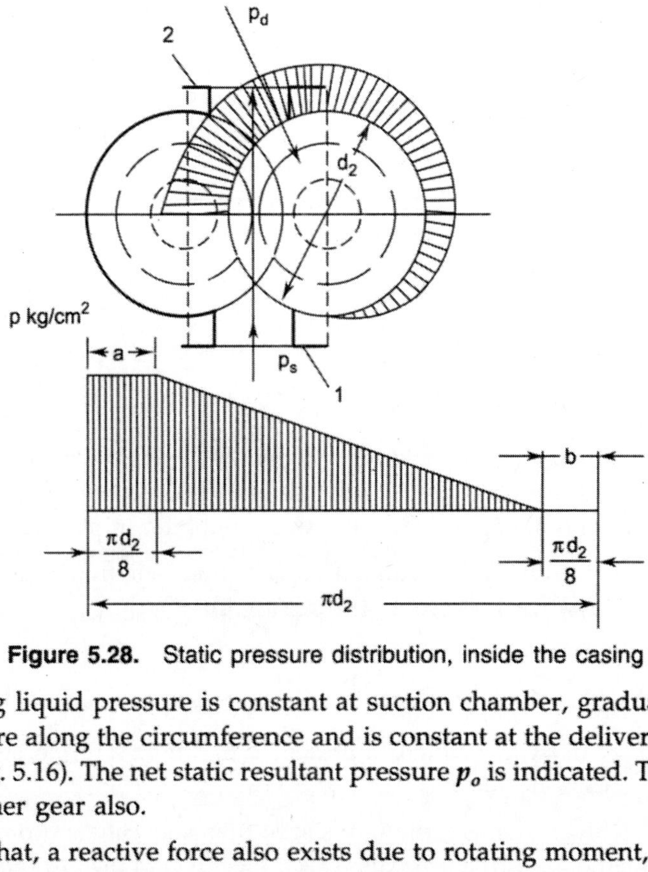

Figure 5.28. Static pressure distribution, inside the casing

The pumping liquid pressure is constant at suction chamber, gradually increases to delivery pressure along the circumference and is constant at the delivery chamber. This is shown in (Fig. 5.16). The net static resultant pressure p_o is indicated. The same pattern exists at the other gear also.

Apart from that, a reactive force also exists due to rotating moment, the direction of which is same as p_o for driver and opposite for the driven. The bearings are subjected to high fatigue force. It is found that, the bearings for driver gear breaks out quite frequently than that of driven gear. Calculation must always be done for forces acting on the driver gear. Based on the experimental results, Prof. E.M.Udhin has given a formula for force P as

$$P = 0.85p\,b\,d_2 \qquad ...(5.55)$$

Because of heavy fatigue force, acting on the shaft and on the driver gear bearings, the operating pumping pressures are always selected considering this fatigue force.

5.17 GEAR PUMP CALCULATIONS

- Gear pumps with two gears are normally used either with the inclined teeth or herring bond type gears. The volumetric efficiency is taken $\eta_v = 0.7$ to 0.95 depending up on the delivery pressure. The flow rate Q_{th} is calculated $Q_a./\eta_v$

- Allowable peripheral speed, u_e is

- $$u_e = \sqrt{\frac{2g\left(p_e - p_{i\min}\right)}{\gamma\left(1-\left(\frac{R_i}{R_e}\right)^2\right)}} = \sqrt{\frac{2g\left(p_e - p_{i\min}\right)}{\gamma\left(1-\left(\frac{Z+2+2\xi_B}{Z-2.6+2\xi_c}\right)^2\right)}} \quad ...(5.56)$$

- Where $p_{min} = \left(p_{vp} + 0.3\right)\frac{kg}{cm^2}$. The velocity u_e for viscosity of liquid ν = 1 to 40cs will be 8 to 4 m sec.
- Number of teeth Z = 8 to 14.
- The correction co-efficient for Ist approximation is taken as $\varepsilon_c = \varepsilon_B = 0.02\ (30 - Z) \approx 0.5$ or $\xi_c = \xi_B = 0.5$. The velocity u_e for visous liquids, v up to 40 ccs will be taken as 8 to 4 m/s
- Module 'm' for the gear is determined from flow rate Q_{th} and velocity u_e

$$Q_{th} = 4\pi r\,(R_e - r)\,bn = \pi m\,(z + 1\varepsilon_b)\,2\,mk_B\,m\,\frac{u_0}{\pi m\left(Z+2\varepsilon_B\right)} \quad ...(5.57)$$

$$m = \sqrt{\frac{Q_p}{2k_B\,u_0}} \text{ and} \quad ...(5.58)$$

$$u_0 = u_B\,\frac{Z+2\xi_B}{Z+2+2\xi_B} \quad ...(5.59)$$

The width of the gear is selected proportional to module as $b = k_B m$ and the co-efficient k_B is related as a function of the delivery pressure, Δp.

Table 5.3

$\Delta_p\ \frac{kg}{cm^2}$	Up to 10	Up to 13	Above 40
K_B	>>13	7 to 9	Up to 6

If speed is known, the module 'm' will be selected as

$$m = \sqrt[3]{\frac{Q_p}{2\pi k_B\left(Z+2\xi_B\right)}} \quad ...(5.60)$$

This value of 'm' is rounded off to the nearest number.
Normal value of 'm' = 3 to 6 mm.

The speed, $$n = \frac{60.u_e}{\pi m\left(Z+2+2\xi_B\right)} \quad ...(5.61)$$

The speed must be slightly changed to suit the nearest prime mover speed. Table 5.3 is used for the calculation of other parameters. The final value of width 'b' is determined from the equation 5.15.

For inclined teeth, u_e is determined and selecting the correction co-efficient 'ξ_s' at the side section and knowing the speed n, the module 'm' is determined as

$$m_s = \frac{60.u_e}{\pi n\left(Z+2+2\xi\right)n} \qquad ...(5.62)$$

And width

$$b = \frac{Q_p}{2\pi m_s^2\left(Z+2\xi_B\right)n} \qquad ...(5.63)$$

The angle of inclination of the teeth 'b_d' is determined by the extended length of cylinder as

$$\tan\beta_d = \frac{\pi m_s}{b}\left(\varepsilon_s - 1\right) \qquad ...(5.64)$$

The normal value of module $m_n = m_s \cos\beta_d$, is rounded off to the nearest standard value. From the known value of m, and the angle β_d, the new value of module for the side m_s, is determined and other parameters of the gear. The flow rate is verified using equation (5.57). The inclined tooth gears Herring bond gears are adopted for higher flow rate of 200 m^3/hr and viscosity $v \approx$ 2000cs, if more vapour and gases are present. The angle 'β' for Herring bond type gears, will be up to 30°. By this, the contact line between the suction chamber and the delivery chamber will be absent.

CHAPTER **6**

SCREW PUMPS

6.1 PRINCIPLES

Screw pump is another type of gear pump where Length / breadth (L/D) ratio is larger. Screw pumps are of the same type as that of gear pump except that in gear pump, liquid is displaced along the circumferential direction whereas in screw pump the liquid is displaced in axial direction.

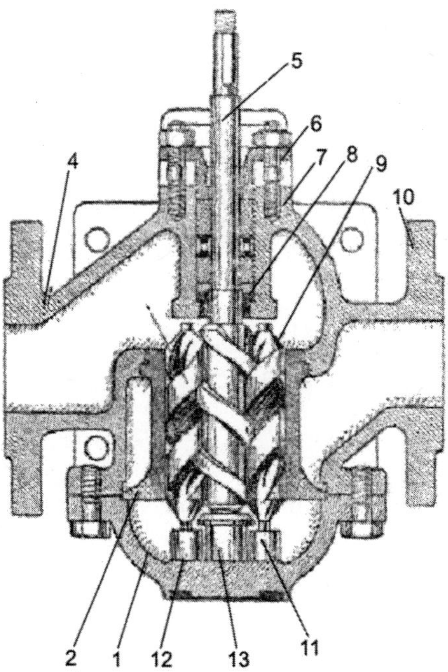

1. Cover with bearing support. 2. Rotor casing. 3 and 9. Driven rotors. 4. Delivery chamber. 5. Driver rotor with shaft. 6. Gland. 7. Casing. 8. Sleeve bearing.10. Suction chamber. 11 and 13. Unloading device for the driven rotors, 12. Unloading device for the driver rotor.

Figure 6.1. Screw pump construction

This is also a static and positive displacement pump. The rotor is in the form of screw. Usually number of rotors will be three or more, the central screw being the driver and other two are the driven screws (figure. 6.1). In some of the constructions a single long screw is also used to transfer solid material such as coal powder, sand etc.

6.2 CLASSIFICATION

Screw pumps can be classified into the following :

1. As per the profile form of the screw, as seen from the side view such as cycloid and special profile.
2. As per the number of screws such as double or triple or many screws.
3. As per the number of parallel flows such as single start and two start.

Screw pumps are used for a flow rate Q from 2 to 500 m^3/hr up to a pressure of 200 kg/cm^2, the speed up to 3000 rpm. Efficiency ranges from 60% to 85%. Viscosity of pumping fluid ranges from 10 to 2200 CS. These pumps can be coupled to a high flow rate unit, Smaller size and dimension, self-priming, can continuously work, more uniform inflow are the major advantages of this pump. Complicated profile is the draw back of the pump, which makes manufacturing more difficult. Screw pumps are operated for a flow rate of 0.5 to 250 m^3/s developing pressure up to 30 to 2000 MWC, and speeds between 300 and 8000 rpm. These pumps are used to pump viscous fluid at different temperatures.

The main advantage of screw pump is

1. simple in construction
2. less weight
3. noiseless operation
4. can be used for pumping liquids of high viscosity
5. can also be used for high speeds of rotation.

6.3 CONSTRUCTION

All the rotors are positioned is one housing with closed running clearance between the rotors and the housing. The radial clearance between the rotor and housing should be between 0.15 mm to 0.25 mm. If the clearance is more than 0.25 mm, the volumetric efficiency reduces considerably, due to leakage through clearance from delivery to suction.

Suction pipe is connected to the suction chamber. The liquid, at the suction chamber, occupies the space between teeth. When the rotor is rotated, this liquid is displaced gradually from suction to delivery chamber through the space between the teeth of both the driver and the driven screws.

It is very important to maintain complete hermetic condition, *i.e.*, air tight condition between delivery and suction chamber. In order to attain this condition, the radial clearance between rotor and housing as well as the gap between the teeth of driver and driven at the time of contact at non-pumping area must be as minimum as possible. The radial clearance between rotor and housing purely depends upon the accuracy in manufacturing, whereas the leakage between the gap between driver and driven depends upon the friction in flow passage as well as the space between two teeth.

It is essential that the length of the rotor must always be many times larger than the pitch of the screw, in order to avoid the direct contact between delivery and suction chambers, through the flow passage, *i.e.*, the space between two teeth of the rotor, especially during idle condition.

Two conditions must be ensured in order that the screws of the two rotors shall always be in contact between them.

1. The contour of teeth of two screws must be in opposite direction one in clockwise and another in anticlockwise direction,
2. The contour of the profile surface normal to axis must be a proper curve, ensuring the laws of contact, obeying Willies theorem. /3/ The profile of the tooth of engaging screws must ensure a continuous kinematic contact along the line of contact. This can be obtained only, if the side surfaces of the two teeth are in continuous contact with each other at all times.

The form of tooth profile for the driver and the driven is determined by the line of contact. Normal cycloid profile or any other evolved profile, where the line of contact is a straight line, will not fulfill the requirement. Only special cycloidal profile can provide complete shape of the tooth profile. /15/ The two meshing profiles must have a contact at two points, where one profile will have a convex shape of curve, while the other profile will have a concave shape. The curvature of concave profile must be greater than the curvature of convex profile. These meshing curves will be epicycloids for convex surface and extended epicycloids for concave surface. The drawn points to form the shape of the profile must be located as one outside the starting circle on the extension of its radius and the second either on this curve (or) inside the circle (figure 6.2). /33/

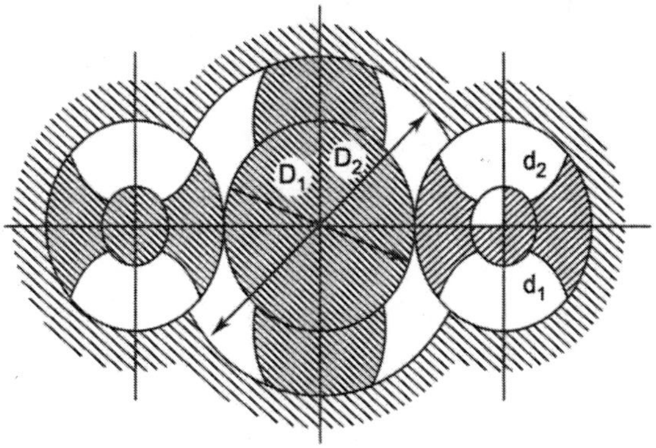

Figure 6.2. Cross-section of a three rotor screw pump

Most commonly adopted screw pump is the three rotor pump; the centre is the driver, while the other two are the driven. The gear ratio is 1:1. Double start screws with cycloid profiles are also adopted.

Independent of the capacity and delivery pressure range, the dimensional ratio between basic dimensions (D_2, D_1, d_2, d_1) remains same. This ensures quick and easy design and manufacture of these pumps. The nominal or reference diameter for a screw pump is the diameter, d_2. All other dimensions are determined as a ratio of this diameter such as

$$D_1 = d_2,\ D_2 = 1.67d_2,\ d_1 = 0.33d_2,\ t = 3.33d_2$$

where,

D_1 = inner diameter of the driver tooth of the screw

D_2 = outer diameter of the driver tooth

d_1 = inner diameter of the driven tooth of the screw

d_2 = outer diameter of the driven tooth

t = pitch between the teeth of the screw (fig. 6.5)

6.4 FORCES AND MOMENTS ACTING ON SCREW PUMP

The torque on the driven screw consists of a force pair G, created due to the pressure p acting on the projected area of the flow passage between teeth in a plane surface, passing through the axis of rotation (figure 6.3). This torque is proportional to the area F_{cv} and F_{cc} of the flow passages. (Fig. 6.4), and is equal to the moment of force G_{rad}, created, due to the delivery pressure p acting on twice the projected area $\mathbf{A_r}$ along the axial plane (the area A_r is taken as twice because of two passages, due to double teeth in the rotors). For the selected dimensions and the clearance passage, the turning moment developed on the driven due to the fluid pressure acting on flow passage area will create opposite moment, on the other side on the surface of the flow passage on area 1,2,3,4 **(A_r)** due to the direct contact through hub. The reaction of the fluid pressure on the area $\mathbf{A_r}$ acts on the hub surface of the driving screw A. If the area of the teeth of the driven is also A_r, then the turning moment on the screw B will be equal to zero and will be transferred to the driver A. Apart from the main torque, an additional frictional moment acts on the driven, due to rotation of the screws, The area A_r must be more than area A_{cv} in order to compensate the additional turning moment created by the friction. there by the driven screw is completely free from the torque. The driven screw rotates purely due to the fluid pressure. Equalization of any extra turning moment acting on the driven can be done only by isolating it from the suction and delivery pressures. This is attained by proper selection of the value a (figure 6.5) for the selected ratio between the constant area A_r and the variable area A_{cv}. The value $a = 0.3\ d_2$ in order to have $A_r > A_{cv}$.

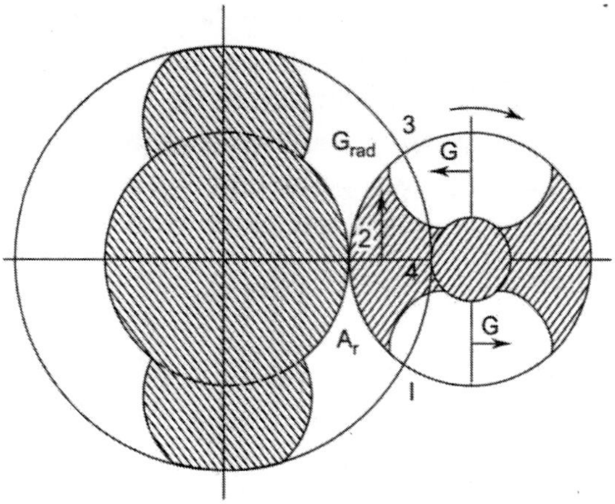

Figure 6.3. For the calculation of torque in screw pump

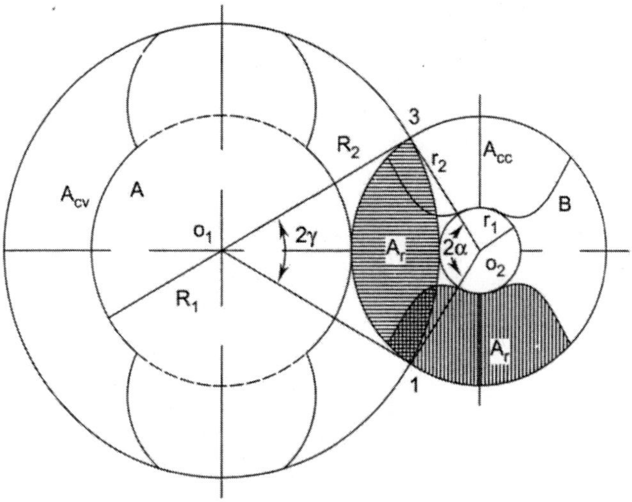

Figure 6.4. Method of balancing the turing moment

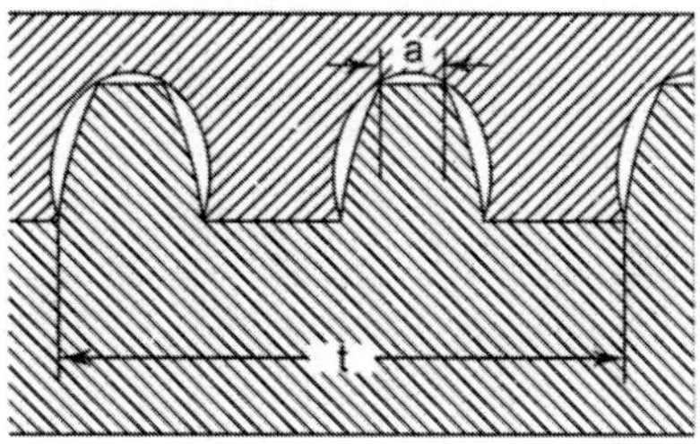

Figure 6.5. Spread out view of teeth engagement

Usually the screw pumps have one driver screw meshing with one or two or more, up to five, driven screws. All these screws are housed in one casing and made to rotate inside casing with very close radial clearance between casing and screws (figure 6.6).

In figure 6.6(*a*), a twin screw with spur gear tooth profile is illustrated. The driving screw 4 has right hand teeth and the driven screw 7 has a left hand teeth. The casing

1 consist of the housing to accommodate the twin rotors with a closed radial clearance. The casing is connected to the suction 3 and delivery 5 chambers through flange connections. Radial thrust pads 2 are provided at the bottom side of the rotors to take care of the axial thrust. Constant uniform and synchronous speed of rotation for the screws are obtained though meshing gear 6. These two gears are connected to two screws of the pump. In figure 6.6.(*b*), a three screw pump with cycloid teeth profile is illustrated.

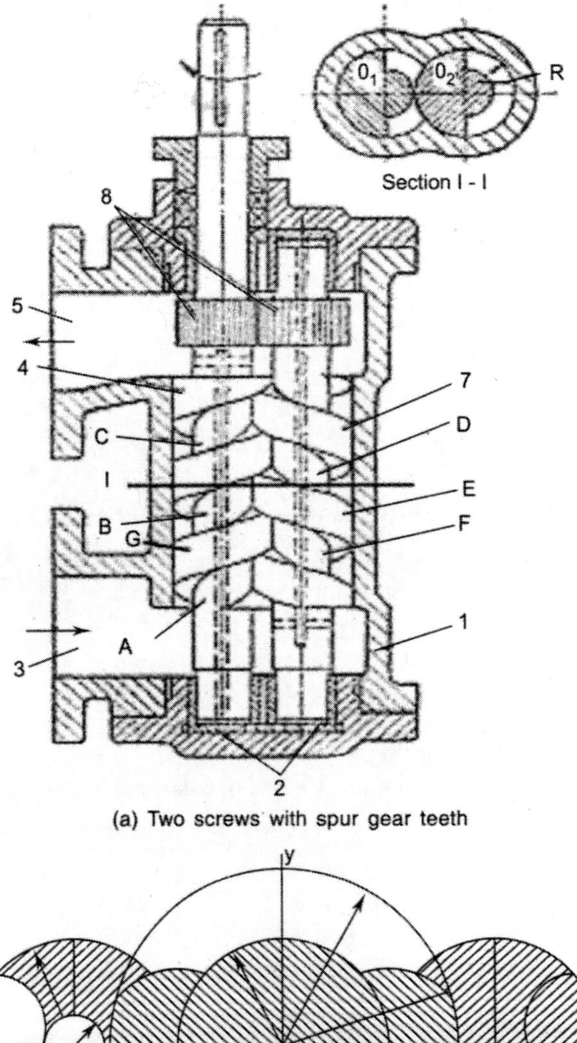

(a) Two screws with spur gear teeth

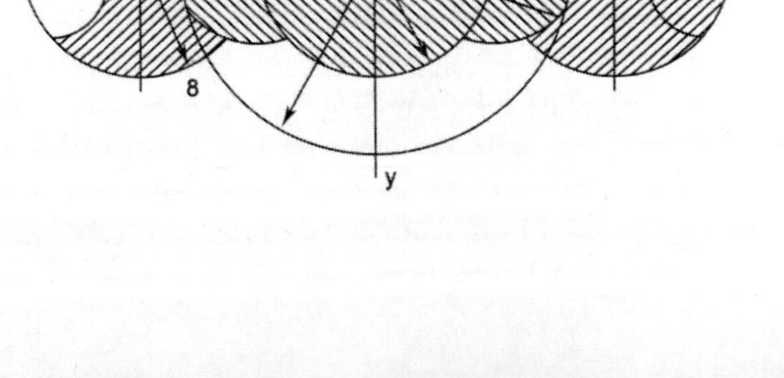

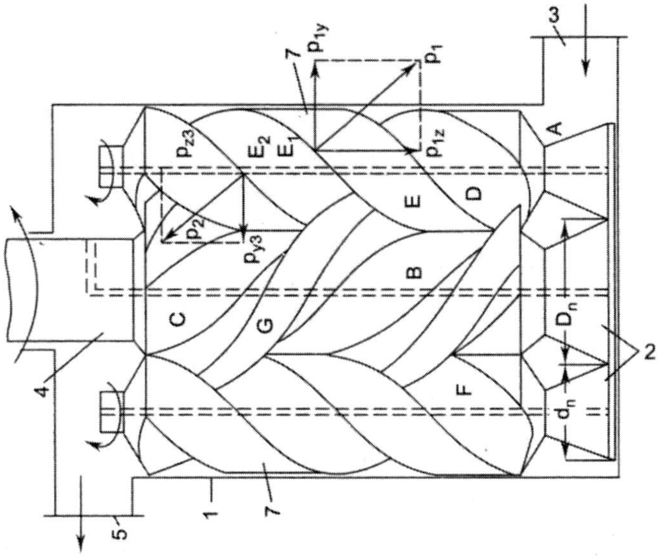

(b) Three screws with cycloid teeth

Figure 6.6.

The constructional details of this pump is same as that of pump in figure 6.6(a), except that there is no meshing gear 6, since this is not necessary. The cross-section of both the screws is also given at the side.

When the suction chamber 3 is filled with the liquid, the space between two teeth of the driver and that of the driven will also be filled with the liquid (point A). When the driver rotates, the liquid in the space A will gradually move from the suction side 3 to the delivery side 5, through the space B, C, F, D of the driver and the driven, due to the driving action of the teeth I and G.

Thus whenever the pump runs, the liquid is displaced through the space between the teeth of the driver and that of the driven from suction to delivery. The energy increase for the pumping liquid takes place due to the constant contact of the teeth with the liquid. The driving force of the teeth is transferred to the liquid due to the contact of teeth with liquid.

Considering figure 6.7, the force acting at any point on the screw teeth can be determined as follows.

Assuming that the point, considered on the driven screw, is isolated, *i.e.*, the driven screw teeth is hermitically sealed, from the delivery pressure p_d, acting on one side and suction pressure p_s, on the other side, the teeth surface is subjected to a net pressure $p = p_d - p_s$. This pressure p uniformly acts on the surface of the screw teeth, on the delivery side of the screw. The elementary force $dP_T = p.dA$ where dA is the elementary area, acting by the screw on the fluid, can be assumed to act at an arbitrary point M, on the pitch circle line EML of the screw surface and normal to the point considered. Taking the point 'M' as origin, the resolved components of the force. P_T along the axial, tangential and radial directions are

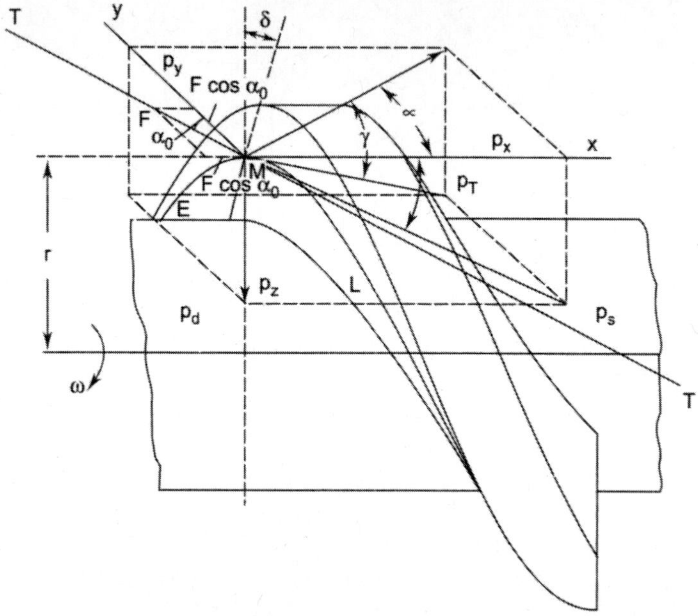

T – T Tangents to the average line at point M on the screw surface

Figure 6.7. Forces and fluid pressure acting on the screw surface

Axial force, $P_x = P_T \cos\gamma \cos\alpha$ ...(6.1)

Tangential force, $P_y = P_T \cos\gamma \sin\alpha$ ...(6.2)

Radial force, $P_Z = P_Y \sin\gamma$ ...(6.3)

When the screw rotates with an angular velocity ω the work done 'W' by the axial force P_x and by the radial force P_z are Zero, The moment M_T, developed by the tangential force P_y will be

$$M_T = P_y.r = P_T \cos\gamma \sin\alpha \quad ...(6.4)$$

Neglecting the losses in the pump, this moment M_T is the moment of the screw acting on the fluid and is equal to the resistive moment, which is created due to the difference between the delivery and suction pressures acting on the tooth surface. Thus, it is found, that certain amount of energy must be spent in order to keep the suction and delivery pressures at its level. Perfect sealing is achieved by providing a continuous contact between the driver and the driven screw, which is obtained by selecting a suitable shape of the tooth surface and suitable number of teeth for the screw, thereby, the resistive moment, due to sliding of driver and driven teeth surface during the rotation of the screws, is made almost to zero.

The friction force 'F' developed due to the contact surface between teeth of the driver and driven at the time of rotation, is acting along the tangential direction on the pitch circle at the point M. This friction force 'F' lies on the surface *x-y* and can be resolved along these directions.

The axial force, $F_x = F \sin\alpha_0$, acting opposite to the axial force P_x and the tangential force $F_y = F \cos\alpha_0$, which gives a moment.

$$M_f = F_y.r \quad ...(6.5)$$

This frictional moment increases the actual power of the prime mover. It is very important to see that the constructional features of the screws do not influence and limit the economy and on the magnitude of the delivery pressure. In order to achieve this, *i.e.*, the leakage between the delivery and suction with minimum mechanical friction losses at the contact point of two teeth, the following rule should be adopted./15,20,29/

In order that the contact is perfect, continuous and tangential to the profile surface ABC, the corresponding arcs of the circle should be d_2 and $5/3\ d_2$, figure 6.6 (*b*).

The length of the screw is always more than the pitch, as otherwise perfect sealing between delivery and suction cannot be achieved. The cross-section of the profile (figure 6.6*b*) should provide perfect contact between teeth, *i.e.*, the cross-section (side view) of the screw (fig 6.6*b*) follows the rules of gear teeth engagement, *i.e.*, the normal at the point of tangential contacts of the side surfaces of the tooth passes through the tangent to the base circle (polus).

The rule of Mantelus is followed for the air tightness *i.e.*, if the number of driven screws is m with Z_2, number of flow passages, then the number of flow passages of the driver Z_1 must be,

For convex profile $\quad Z_1 = m\,(Z_2 - 1)$

For concave profile $\quad Z_1 = m\,(Z_2 + 1)$ ...(6.6)

If the profile of the tooth is trapezoidal, leak proof effect between the teeth, *i.e.*, perfect contact between the driver and the driven at all times cannot be achieved, which increases the leakage flow from delivery to suction side. Although the clearance can be reduced by considerably increasing the pitch *i.e.*, increase in length of the screw, the overall size of the pump is increased, which is not desirable.

Required length of the screw can be obtained by reducing the pitch of the screw. This can be achieved by reducing the pressure angle of the tooth. Smaller angles should not be selected because of self breaking effect between the screws. Under such conditions, it is necessary to provide a gear wheel connection between driver and driven in order to obtain a uniform rotation (figure 6.6(*a*)), since the power transmission from the driver direct to the driven is not possible.

All types of Mantelus screws provide an airtight continuous contact all times of rotation between driver surface and the driven surface of the tooth. Hence the reduction in the angle of rise is not at all considered. The pressure angle of the screw "β" in these pumps lies between 26° and 46° which eliminates the self-breaking effect and corresponding elimination of gear transmission. The power is transmitted from the driver to the driven directly. However the length of the screw must be increased, if the delivery pressure is increased in these hermitically sealed pumps as well as in pumps of non-sealed type, in order to keep volumetric efficiency same. However the volumetric efficiency is maintained in hermitically sealed units, but not in non-sealed units, even when the length is increased. That's why, Montelus profile is adopted for medium and high pressure units. One such screw pump with cycloid profile of teeth with all details is illustrated in figure 6.8.

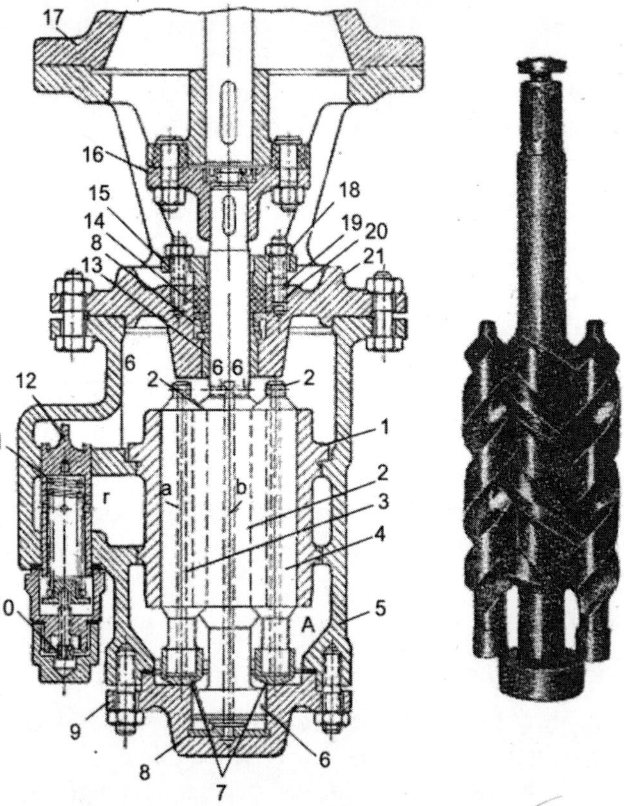

Figure 6.8. Cross-section of a screw pump

These pumps are used for pumping and supply of lubricating oil for various machines. Over and above the normal construction as mentioned earlier, additional features provided in this screw pump, figure 6.8, are the axial balancing devices and high pressure protection. The axial thrust acting on the driver is balanced by the balancing disc 6, thereby heavy axial thrust does not act on the axial thrust pad, provided at the bottom. The high pressure oil, from the delivery side reaches through two radial channels and one axial hole to the bronze bearing of the driver 8 kept at the bottom. For the balancing of driven screws oil from the delivery side passes through two radial and one axial hole to the axial thrust pad, kept at the bottom, thereby axial thrust pad is subjected to a very low pressure only. The holes are designed to balance the axial thrust.

A safety valve is provided to protect the pump against the high pressure development. At high pressures, approaching the yield strength of the material of the pump, the safety valve automatically opens and connects the delivery chamber to the suction chamber. The high pressure oil from the delivery is directed to the suction chamber through the safety valve, thereby high pressure is released. The spring of the safety valve is adjusted to the high pressure. The valve automatically closes, when the delivery pressure gets reduced to the normal value.

6.5 FLOW RATE

The volume displaced, during one revolution of the screw, will be equal to the space between screws for one circular pitch distance. Actual area A is the area of cross-section filled between two successive teeth and the casing multiplied by the number of teeth. This is equal to the difference between the area of casing 'A_c' and the area occupied by the teeth of both the driver and the driven, A_s. = A = $A_c - A_s$. The Radial clearance between the casing and the addendum of both teeth is neglected.

For the epicycloids profiles, selected, a_1b_1 and a_2b_2, the area A_s can be determined; so also the area of the housing, A_c, where the screws are located.

Based on the experimental results, the area A can be expressed in terms of nominal diameter d_2 as

$$A = K_1 d_2^2 \quad ...(6.7)$$

where K_1 is the constant, depending upon the tooth profile such as trapezoidal or epicycloid or Mantelus tooth. $K_1 \approx 1.25$ for theoretical profile, and ≈ 1.24 for corrected profile. The flow velocity C can be expressed from the speed 'n' rpm and the pitch $t = \frac{10}{3} d_H$

$$C = \frac{tn}{60} \text{ m/sec} = \frac{10}{3} d \frac{n}{60} \text{ m/sec} \quad ...(6.8)$$

The flow rate

$$Q_{th} = A \times C = 1.24\, d_H^2 \frac{10}{3} d_H \frac{n}{60}$$

$$= 0.0688 d_H^3\, n \frac{m^3}{\text{sec}}$$

$$= 247.8 d_H^3\, n \frac{m^3}{hr} \quad ...(6.9)$$

For a two screw pump with trapezoidal spur gear profile, the same procedure can be followed.

$$A = A_C - A_s = \frac{3}{16}\pi\left(d_2^2 - d_1^2\right)$$

where d_2 = addendum diameter

d_1 = base diameter

Theoretical Flow rate, $$Q_{th} = 11.25\,\pi\,(D^2 - d^2)\, tn \frac{m^3}{hr} \quad ...(6.10)$$

Knowing the clearance between the screws and casing, actual discharge, Q_a will be less by an amount of leakage flow. If the volumetric efficiency is, η_v, $Q_a = \eta_v\, Q_{th}$.

For screw pump η_v ranges between 0.7 and 0.98. Higher values are taken when $Q_{th} > 15$ lit/sec, and delivery pressure $p_d > 15$ ata.

For the given quantity Q_a and known volumetric efficiency, η_v the nominal diameter d_H can be calculated using equation.

$$d_2 = \sqrt[3]{\frac{Q_{th}}{0.0688n}}, m \qquad ...(6.11)$$

From the known value of d_2, all other dimensions can be determined.

6.6 PRESSURE ON THE SCREW

The fluid forces on the surface of the screw are the axial force F_z, radial force F_r and the moment M_z. The moment M_z due to the fluid pressure p on screw surface with respect to the axis of rotation z is equal to

$$M_z = \int_A rpd\,A\cos(u, n) \qquad ...(6.12)$$

where dA is the elementary screw surface.

r is the distance from the area dA up to the axis z.

(u, n) the angle between the external normal n to the area dA and the peripheral velocity, u.

The surface area of the screw A is the sum of the cylindrical area A_{cy}, side area A_s and screw area A_{sc}, *i.e.*, $A = A_{cy} + A_s + A_{sc}$. Neglecting friction, the pressure on the cylindrical and side surfaces do not give any moment M_z since the forces, either passes through the axis z or parallel to it. Hence the moment M_z is 0, due to the forces on the profile passage surfaces. Pressure changes along the flow path, from suction to the delivery, correspondingly on the profile surface. At any section of the flow path between inlet and outlet, the pressure along the circumference is same. Moment M_z for the driven shaft can be calculated as follows: The circumferential pressures at the intermediate portions of the screw are equal and opposite to each other and hence cancelled. The pressure at one end of the profile is suction pressure and at the other end is the delivery pressure.The value $dA_u = dA_B \cos(u, n)$ is the projection of elemental surface dA_B on the plane perpendicular to the screw velocity, u.

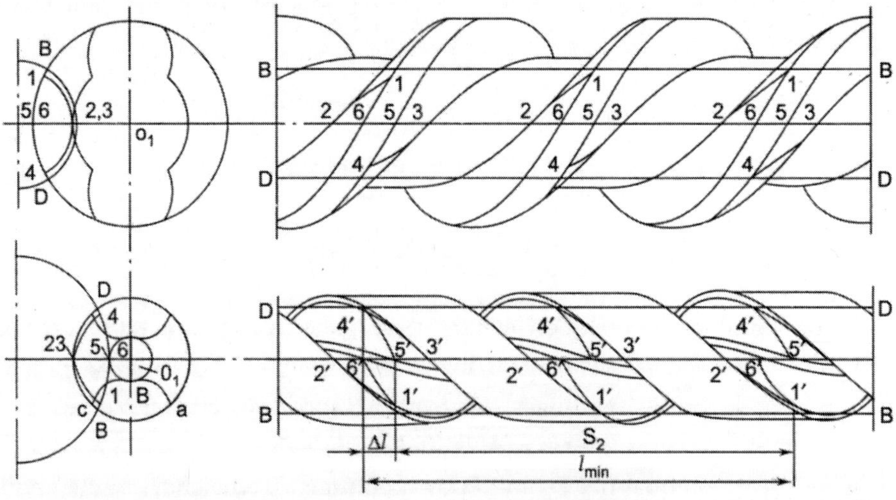

Figure 6.9. Details of the driver and the driven profile indicating the line of contacts

For a screw surface

$$\tan \beta = \frac{dA_u}{dA_z} = \frac{s_2}{2\pi r}$$

where β is the pressure angle of the screw and dA_z is the projected area of dA_B normal to z axis. Equation (6.12) can be written as

$$M_z = \int_A rpdA \cos(u, n) = \int_A rpd\, A_u = \int_A \frac{ps_2}{2\pi} dA_z \qquad \text{...(6.12(a))}$$

Moment M_z, due to the fluid pressure on the surface of one passage of the right portion of the driven screw, O_2, will be

$$M_{zi} = \frac{p_d S_2}{2\pi} \int_{A_{zz}} dA_z = \frac{p_d S_2}{2\pi}(A_R - A_L) \qquad \text{...(6.13)}$$

where p_d is the delivery pressure area A_R = area *baB16b*, the projection of the side area at the right hand side of the passage and A_L = area *bc245b* is the same at the left hand side (Fig. 6.9).

The value $\quad (A_R - A_L)$ = area A_1 (= *bacb*) – area A_2 (= 12451)

So,

$$M_{zi} = \frac{p_d S_2}{2\pi}(A_1 - A_2) \qquad \text{...(6.14)}$$

If there are z_2 flow passages, in the driven screw, the total moment M_{zd} due to fluid pressure on the passage surface z_2 of the profile at the delivery side is

$$M_{zd} = z_2 M_{zi} = z_2 \frac{p_B S_2}{2\pi}(A_1 - A_2) \qquad \text{...(6.15)}$$

The moment M_{zR}, due to the fluid pressure at the end of the flow passage, connected to the suction chamber will be

$$M_{zR} = -Z_2 \frac{p_d S_2}{2\pi}(A_1 - A_2) \qquad \text{...(6.16)}$$

The total moment

$$M_z = M_{zd} + M_{zi} = \frac{Z_2 S_2}{2\pi}(p_d - p_s)(A_1 - A_2) \qquad \text{...(6.17)}$$

where p_d and p_s are the delivery and suction pressures respectively. Usually $A_1 = A_2$, Then the moment M_z for the driven is equal to zero. In practice for better performance A_2 = (1.02 to 1.03) A_1. Under this condition axial force acts on the driver screw and will add along with friction force acting on the driven screw. Since no force is required to rotate the driven from the driver, the wear is reduced and hence almost no friction prevails.

The axial force $F_z = Z_2(p_d - p_B)(A_1 - A_2)$ acting on the driven screw is directed from delivery side of the casing to the suction side.

6.7 AXIAL FORCE AND BALANCING DEVICES

Axial force is the net pressure difference between the delivery pressure p_d and suction pressure p_s, and it acts on the projected surface of the tooth of the screw pump, the direction of which is from the delivery to the suction and is in axial direction. Apart from that, the friction moment developed by the driver and driven screws enclosed in the casing also produces an axial force. This force will be acting towards suction side by the driven screw and towards delivery side by the driver screw. If the pressure drop between the delivery and section sides of the screw teeth is very small, it can be neglected (Figure 6.7).

This axial force is normally taken care of by the axial pads kept at the bottom of the screw or at the suction end of the screw. Higher axial force is balanced by providing high pressure at the axial pad, there by equivalent and opposite force is created to neutralize the axial force. A plunger is used to provide high pressure, to balance the axial thrust. For a three screw pump, the diameter of the plunger $D_p = 1.82\ d_2$ and $d_p = 0.67\ d_2$, where D_p is the plunger diameter of the driver and d_p is the plunger diameter of the driven.

Referring to figure 6.7, the tangential force p_y provides the hydraulic friction moment. The axial force p_x is taken care of by the axial thrust pad (or) bearing. The radial pressure, p_z acting on the driver is uniformly distributed relatively to axis Y-y and Z-z there by the radial force is zero. On the surface of the driven screw also the radial pressure acts, but in opposite direction, due to the opposite direction of rotation with respect to the driver. Since the driven screws are mounted eccentrically with respect to the shaft axis, symmetry cannot be attained, as a result of which radial force is not equal to zero as attained in the driver. It is calculated by an empirical formula.

$$P_z = 1.40\ pd_2^2. \qquad ...(6.18)$$

The direction of this force is parallel to the peripheral velocity direction of the driver and the tendency is to reduce the radial clearance between the driven and the casing at the outer side.

Referring to the points E_1 and E_2 in the figure 6.6(*b*), and the point M in the figure 6.7. Assuming that, the neutralizing fluid forces P_1 and P_2 acting on the screws, act on the plane parallel to the axis passing through the centre O_2, but separated at a distance (see Fig. 6.6(*b*)), neglecting the force P_z, the forces P_{1y} and P_{2y} provide opposite and equal rotating moment. Based on the resultant moment from these two moments, the driven will rotate along the direction indicated in the figure 6.6(*a*) or in the opposite direction as the case may be. In both cases mechanical friction between the driver and the driven is developed. In order to reduce the friction on the driven screw, developed due to the hydraulic moment, an equal and opposite force P_{2y} is applied so that the driven screw can run without any external moment arise due to the tangential force. This is achieved by the system shown in the figure 6.6(*b*). More over mechanical losses are difficult to calculate. In order that the screws work without any mechanical friction on the driven, the moment applied to the pump should be equal to the frictional moment arising due to mechanical and hydraulic friction, acting on both the driver M_1 and the driven M_2.

$$M_1 = M_{1hyd} + M_{1mech}$$

$$M_2 = M_{2hyd} + M_{2mech}$$

So $$M_T = M_1 + 2M_2$$

If N_T is the power in (kW), Q_{th} is the quantity in (m^3/sec), ω (= $2\pi n$) the angular velocity in rad/sec and $p = p_d - p_s$, the pressure difference between delivery and section, the total moment MT will be

$$M_T = M_1 + 2M_2 = \frac{N_m}{\omega} = \frac{pQ_{th}}{\omega} = \frac{pQ_{th}}{2\pi n} \qquad ...(6.19)$$

Axial component of hydraulic pressure on the screw and on the side surface of the driver p_x and driven p_{1x} are directed towards the delivery side.

For the driver $$P_x = 2.5\ P\ d_2^2 \qquad ...(6.20(a))$$

For the driven $$P_{1x} = 0.4\ d_2^2 \qquad ...(6.20(b))$$

This force is taken by the axial bearing (or) pads, but can be reduced by providing hydraulic unloading screws as shown in figure 6.8. The unloading device consists of the system where in the delivery pressure is brought to the bearing area as given by dotted line in figure 6.8.

In order to equalize the axial force 'P_x' and 'P_{1x}' corresponding diameters of the surface figures 6.6(*a*) and (*b*) should be

$$\left.\begin{aligned} D_n &= (1.4 \text{ to } 1.5)\ d_2 \\ d_n &= 0.67\ d_2 \end{aligned}\right\} \qquad ...(6.21)$$

Sometimes, for axial balancing, double sided suction entry to the pump is constructed but the construction of the pump is complicated.

6.8 SUCTION CONDITIONS

Allowable suction head is determined by the loss of head 'Δh' taking place in suction side of the pump. In figure 6.10, a scheme for a suction condition for a screw pump is depicted./32/

Under normal conditions, the total energy at section I-I of the suction pipe, (figure 6.10),

$E_1 = \frac{p_1}{\gamma} + \frac{C_1^2}{2g} + Z_1 + \Sigma h_{fs}$, taking into consideration, the hydraulic losses between section I-I and A-A. At D-D, the starting point of the fluid in contact with the teeth, the suction chamber pressure p_s must be > p_{vp}, the vapour pressure of the pumping liquid at the pumping temperature.

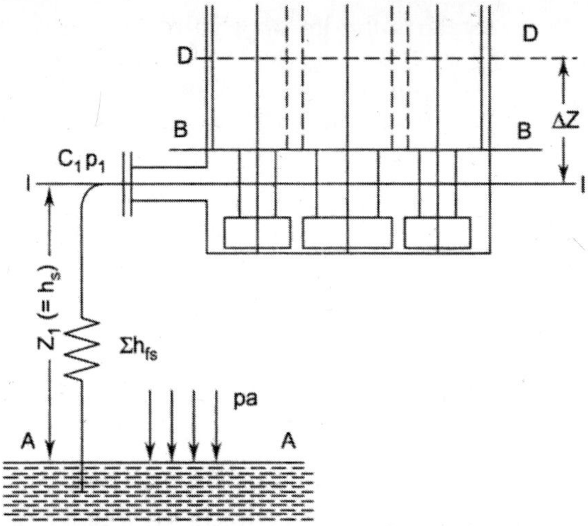

Figure 6.10. Suction conditions in screw pumps

At suction chamber, in a screw pump, periodical closing and opening of the flow passages take place. Referring to the figure 6.6, the flow passage F is in the opened position and is filled with the liquid, but the flow passage A has already been filled and started leaving the suction area. The periodical change of closing and opening of the suction area produces an unsteady flow which takes away extra energy loss Δh, which is the hydraulic energy loss. This hydraulic loss, Δh consist of

(*a*) Sudden expansion of flow passage.

(*b*) Change in the flow direction from the suction pipe to section chamber area.

(c) Shock losses, due to the fluid flow entering the screw, while filling up the passage.

(*d*) Energy loss due to the inertia when the fluid is transferred from the suction chamber to the flow passage between screws (between section B-B and D-D (figure 6.10).

Cavitation can occur at D-D, which corresponds to the fully opened flow passage between two screws. For cavitation free operation, the pressure $p_s = p_{ax}$ should be above vapour pressure p_{vp}, where p_{ax} and C_{ax} are the pressure and the axial velocity at D-D

Neglecting, the height ΔZ, and taking the loss as Δh, the Bernoulli's equation between I-I and D-D will be

$$\frac{p_1}{\gamma} + \frac{C_1^2}{2g} = \frac{p_{ax}}{\gamma} + \frac{C_{ax_s}^2}{2g} + \Delta h \qquad ...(6.22)$$

Where $$\Delta h = \zeta \frac{C_s^2}{2g}$$

Based on the experimental results $\zeta = \frac{160.7}{R_e^{0.714}}$, where Re, Reynolds's number $\frac{4C_s R}{\gamma}$,

where R is the hydraulic radius $\approx 0.375 \frac{A}{\pi d_2}$ for a three screws with epicycloids surface,

A is the actual area of cross-section, and d_2 is the nominal diameter (or) reference diameter. The co-efficient ζ does not influence the screw types

$$\frac{p_a}{\gamma} = \frac{p_1}{\gamma} + \frac{C_1^2}{2g} + Z_1 + \sum h_1$$

$$\frac{p_{ax}}{\gamma} = \frac{C_{ax}^2}{2g} + \Delta h + Z_1 + \sum h_1 \qquad ...(6.23)$$

Taking $p_s = p_{vp}$ and $H_{\text{allow. vacuum}} = H_{\text{max Vac}} - \Delta h$, and neglecting Δh, $H_{\text{allow. vacuum}} \approx$ 9*mwc*. Due to complicated nature of fluid flow condition in screw pump, the value □h is determined only by experiments. Figure 6.11(*a*) and (*b*) illustrate the cavitation characteristics for a high viscosity and a low viscosity fluids.

Speed, suction conditions and flow rate are interrelated. The speed should be such that the liquid fills up the complete flow passage so as to give a quantity Q, without any cavitation.

Axial velocity $C_{ax} = \frac{Q_{th}}{A} = \frac{Q_{th}}{k_1 d_2^2}$,Where $A = k_1 d_2^2$ the working area of the screw. k_1 is the flow coefficient =1.24, for triple screw epicycloid tooth. Sometimes Q_{th} is also written as $Q_{th} = k_2 d_2^2 n$ where n is the speed. In that case $k_2 = 0.069$.

Resolving the above equations the axial velocity Cs can be written as

$$C_{ax} = \frac{\sqrt[3]{\left(k_2^2 Q_{th}\, n^2\right)}}{k_1}$$

$$C_{ax} = \sqrt{\left(\frac{2g\left(\frac{p_a - p_{vp}}{\gamma}\right) - Z_2 - \sum h_f}{1+\zeta}\right)} \qquad ...(6.24)$$

Since for cavitation free operation $p_s \geq p_{vp}$, substituting this condition in the above two equations and simplying, the condition for maximum allowable speed n_{max} is

$$n_{max} = \left[\frac{\sqrt[4]{(2g)^3}}{k_2}\right]\sqrt{k_1^3}\,\frac{\sqrt[4]{\left(\frac{p_a - p_{vp}}{\gamma} - Z_2 - \sum h_f\right)^3}}{\sqrt[4]{(1+\zeta)^3}\,\sqrt{Q_{th}}} \qquad ...(6.25)$$

Since the values for k_1 and k_2, for cycloid profile of Montelus is known, substituting these values in the above equation, n_{max} allowable will be

$$n_{max} \leq 185 = \frac{\sqrt[4]{\left(\frac{p_a - p_{vp}}{\gamma} - Z_2 - \sum h_f\right)^5}}{\sqrt[4]{(1+\zeta)^3}\,\sqrt{Q_{th}}} \qquad ...(6.26)$$

Where Q_{th} is in m^3/sec. Equation 6.25 can be applied to all screw pumps. For other types of epicyclic profiles the numerical value is suitably selected and equation 6.26 is applied.

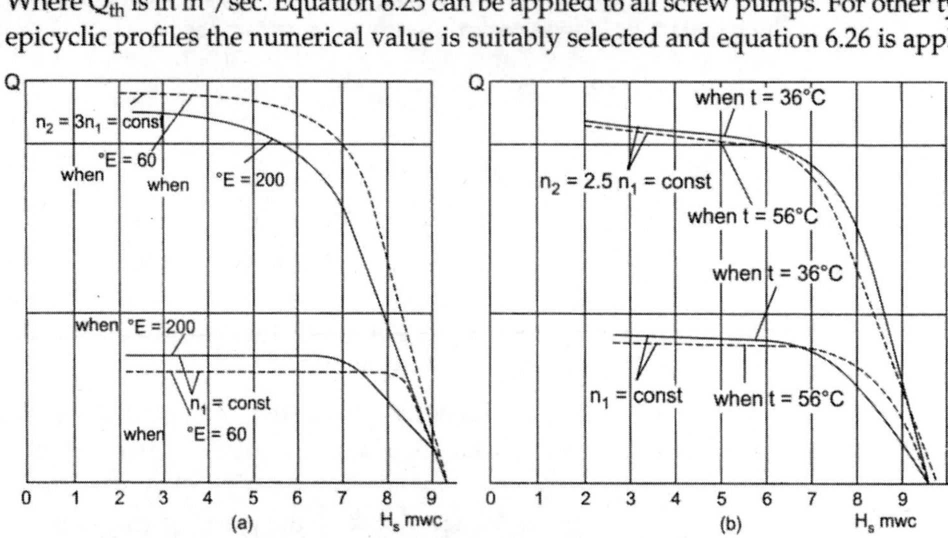

Figure 6.11. Figure (a) and (b) Cavitational characteristics of screw pump at different speeds and viscosities

6.9 SPEED OF SCREW PUMP

In order to achieve a compact size and dimensions of the pump, the speed of the screw should be as high as possible. However, increase of speed necessitates increasing the flow velocity of pumping fluid at suction, in order that the fluid completely fills the suction chamber and also the space between two successive teeth, in order to obtain maximum flow. But high velocity of the fluid in suction chamber may develop cavitation. Correspondingly the flow rate also reduces due to flow separations. Hence optimum speed should be selected based on the cavitation effect.

Flow velocity,

$$C = \frac{tn}{60} = \frac{10}{3} d_d \frac{n}{60} = \frac{d_H n}{18} \text{ or } d_d = \frac{18C}{n}$$

Flow rate, $Q_a = \eta_v \, 0.0668 \, n d_d^3$

Substituting the value of d_H from the previous equation,

$$Q_a = \eta_v \, 0.0688 \, n \times \left(\frac{18C}{n}\right)^3 = 401.76\eta_v \times \frac{C^3}{n^2}$$

$$n = \sqrt{\frac{401.76 \times \eta_v \times C^3}{Q_a}} = 20.044\sqrt{\frac{\eta_v C^3}{Q_a}} \quad ...(6.27)$$

Actual flow velocity, $C_a = \eta_c C_{th}$, depends upon the viscosity of the pumping liquid at the given temperature and suction height.

Figure 6.12 gives the experimental results between actual velocity and kinematic viscosity for the pumping fluid. The above equation can be used to determine the speed of the pump./32/

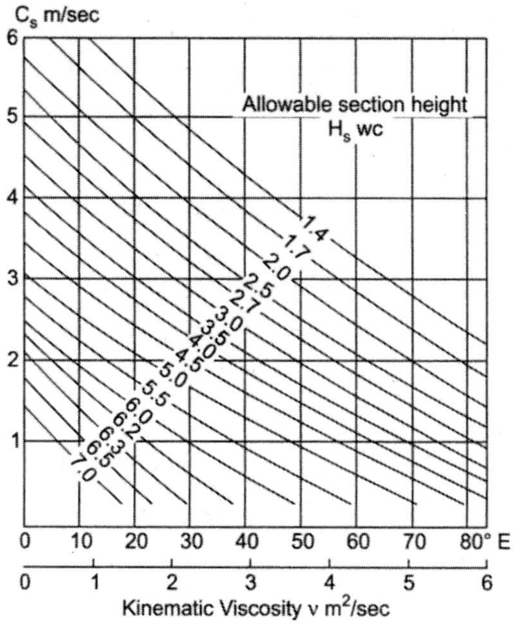

Figure 6.12. Determination of actual velocity for the screw based on kinematic viscosity of the liquid

6.10 LOSSES, AND EFFICIENCY

Volumetric and Mechanical losses are the main losses in screw pumps. Hydraulic losses are negligible, since the flow velocities in flow passage of the pump are low.

6.10.1 Volumetric Losses

Leakage takes place;

1. Through the clearance between rotors and the housing, and
2. Through the radial clearance between teeth of the driver and driven, at the non-pumping area. Total leakage quantity ΔQ consist of the leakage flow between rotor and casing ΔQ_{sc} from delivery to suction and the quantity reduction due to non-filling of the flow passage ΔQ_v.

$$\Delta Q = \Delta Q_{sc} + \Delta Q_v$$

The leakage flow Q_{sc} can be calculated by using the law

$$\Delta Q_c = \mu A \sqrt{2g \frac{p_d - p_s}{\gamma}} \qquad ...(6.28)$$

where 'μ' is the flow coefficient for the clearance which depends upon the construction of pump, pitch and the property of liquid, A, is the clearance area, and $\frac{p_d - p_s}{\gamma}$ is the static head.

In order to attain higher volumetric efficiency the clearance for high pressure pumps will be 0.1 mm, since area reduction considerably reduces the volumetric losses. Also increasing, the length of screw reduces the losses. Economical length,

$$L_{hub} = L_{screw} = Z\frac{10}{3}d_2$$

where Z is the number of pitches of the screw on the working length of the screw. Z = 1 to 1.5 for H < 200, Z = 3 for H = 150 to 600 and Z = 5 for H = 600 to 1500 cap.

The volumetric efficiency $\eta_v = \eta_c \eta_v$. Normal value of η_v = 0.6 to 0.95

Figure 6.13 gives the variations of volumetric efficiency with respect to the flow rate. In this the figure graph η_v *vs* %Q, Curve OAB refers to the epicyclic profile (Mantelus) and curve ODE refers to the spur gear. /32/

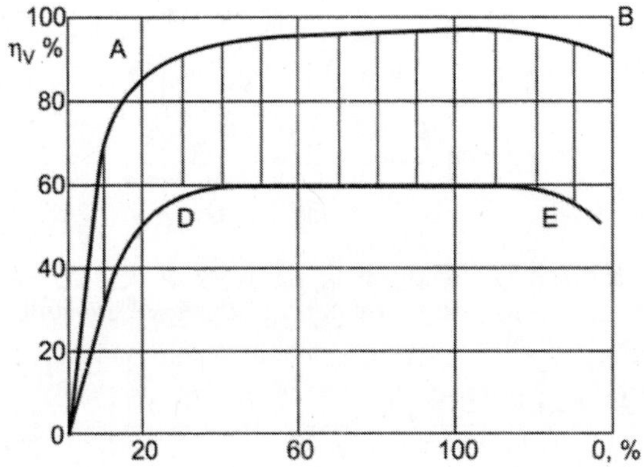

Figure 6.13. Volumetric efficiency as a function of % Q_{opt}

6.10.2 Mechanical Losses

Mechanical losses ΔN_m consist of the losses in the bearings and losses due to moving continuous contact (rubbing) of driver with driven. It is proportional to the speed. Normal values of mechanical efficiency range from η_m = 0.85 to 0.95, the higher value for epicycle gears and lower value for spur gears.

6.10.3 Overall Efficiency

Overall efficiency, $$\eta = \eta_v \eta_m = \frac{pQ}{N}$$

where $p = p_d - p_s$, the difference in pressure between delivery and section and N the power input.

The variation of $\eta = f(p_d\ \%)$ is illustrated in the graph, figure 6.14.

In figure (6.14) lower value, curve OEDM refers to non-sealed *i.e.*, plain air tight units and curve OABC refers to the (air tight) selected units the drop in the curve, *i.e.*, curve BC and .,DM indicates the effect of the relief valve. The change of efficiency with respect to the quality (Q %) is also in the same pattern as that of the delivery pressure (pd %).

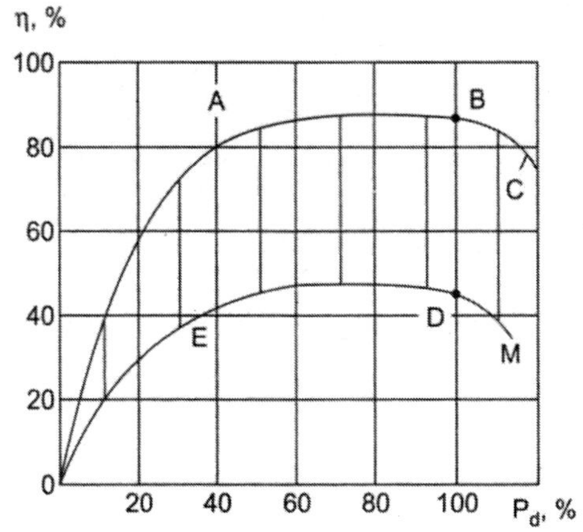

Figure 6.14. Overall efficiency of screw pump as a function of delivery pressure

6.11 PERFORMANCE OF SCREW PUMPS

The performance of screw pump is given by a graph of flow rate Q, power N and efficiency η as a function of differential pressure p or total Head H. under constant speed of rotation *i.e.*, Q, N, $\eta = f(p \text{ or } H, n)$.

Normally the suction pressure will be zero and hence total pressure $p = p_d + p_s = p_d$, the delivery pressure and the total head will be delivery head. (Figure 6.15)

Flow decreases when pressure p_d is increased, due to increased leakage flow, through the clearance. The leakage flow 'q' will be difference between the quantity at zero pressure (p_0) and the instantaneous pressure (p_i) *i.e.*, $Q_{act} = Q_{po} - Q_{pi}$.

Power N increases, when delivery pressure is increased, almost in the form of a straight line. Power N at Q = 0 will be the power spent due to mechanical losses and is independent of the pressure p. So that $N_{act} = N_{spent} - N_{mech\ loss}$.

Universal characteristics or the working characteristics of the pump represents the variation between flow rate and delivery pressure $Q = f(p_d)$ for different speeds, over which equal efficiency (η) curves are drawn. (Figure 6.16)

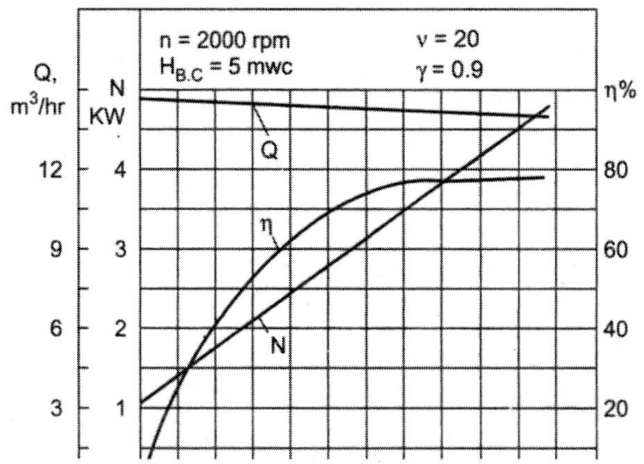

Figure 6.15. Performance of a three screw pump in Q, N, η vs p_d

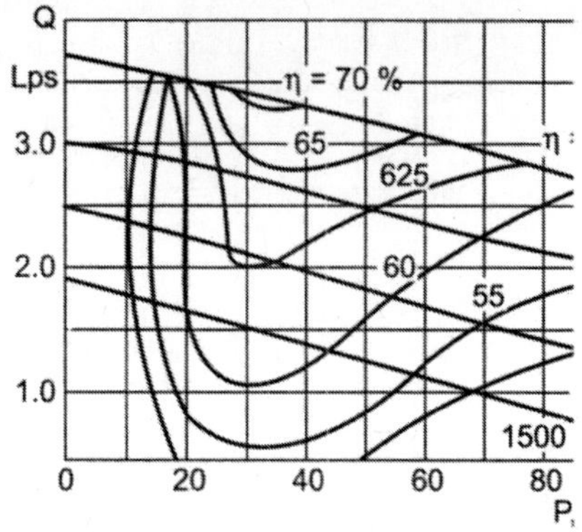

Figure 6.16. Universal characteristics of a screw pump

Cavitation characteristics are drawn between power (N), quantity (Q) and efficiency (η) as a function of suction head (H_s) or absolute pressure $p_B = P_{atm} - \gamma H_s$. The sudden drop in flow rate and in power for the prevailing suction head (H_s) is the limiting condition.

A typical performance of screw pump in (% H NS % Q) format for different speeds and for different viscosities is illustrated in figure 6.17.

In all positive displacement pumps, the flow rate **Q** and the pressure head (H) are independent. The drop in quantity, when pressure is increased is due to the increased leakage flow, which depends upon the speed, viscosity and value of clearance. Figure 6.18 illustrates the change in the quantity Q% as a function of speed (%*n*) for different temperatures of the liquid. Also figure (6.17) gives the effect of power consumption (N) as a function of total head (H), when viscosity is changed.

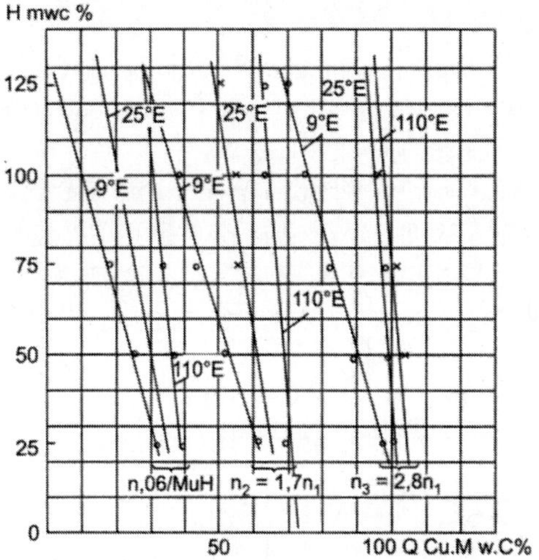

Figure 6.17. H-Q curve for different speeds and the different viscosities for a screw pump

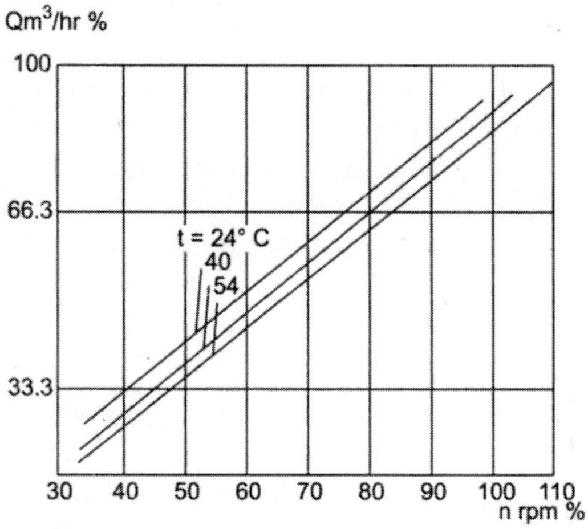

Figure 6.18. Change of Q% as a function of speed n% for a screw pump

6.12 SCREW GEOMETRY

The view perpendicular to the axis of the screw gives the tooth formation. When the screws move at equal speed the three dimensional surface of the screw engaged is obtained which represents the flow passage. The length of passage S with the number of such passages, which is equal to the number of teeth, forms the screw. The pitch t is the axial distance (parallel to axis) between the corresponding point of the tooth, which must be same for both the screws $t_1 = t_2$. So, the passage length $S = Z \times t$.

So
$$t_1 = t_2, = \frac{S_1}{Z_1} = \frac{S_2}{Z_2} \text{ or } \frac{S_1}{S_2} = \frac{Z_1}{Z_2}$$

The screws when engaged will run in opposite direction. Each screw represents like infinite number of teeth with infinitesimally smaller thickness, rotating one relative to each other at infinitesimally smaller angle, with firm engagement between them. The interactive movement the moment can be considered as infinitesimal number of teeth. That's why when number of teeth is less such as 2 or 3, a continuous transfer of rotation from driver to the driven without any special gear connection is obtained. The profiling of the screw surface is carried out as per following requirements. /33/

1. The suction chamber and the delivery chamber must be separated with air tight space between them.
2. The working section of the screw for the given size of pump must be smaller.
3. The driven gear must be unloaded from the rotating moment. It should be free and should rotate only by the driver.
4. The screws must possess sufficient strength.
5. The fluid must pass through the space between gears uniformly from suction to delivery.
6. The wear of gear tooth profile must be minimum.

Leakage flow takes place from delivery to suction through the radial clearance between the gear and casing as well as through the clearance between two gears. The radial clearance between casing and gear is kept as minimum as possible in order to achieve high mechanical efficiency.

In order to avoid continuous flow between the delivery passage and the space between two gear at the time of engagement of two gears *i.e.*, during return, the following conditions are to be stipulated.

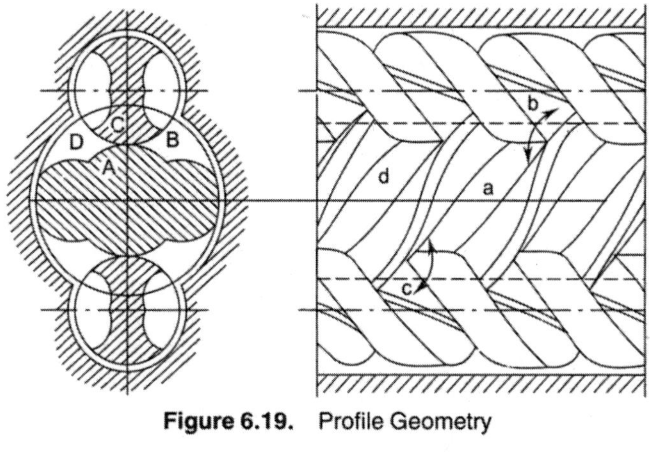

Figure 6.19. Profile Geometry

1. The tangent drawn to the screws must be a continuous line and must pass through the points B and D, the section of external circumference of the screws (figure 6.19). It will not have any contact with the space '*a*' and '*d*' of the screws.
2. The line of engagement of the screws must be a closed one, without any gap or clearance, since this form the projection of tangent line of the screws on the side view (perpendicular to axis). From all the points on the line of engagement passing through points A, B, C, D. A line is obtained, which is the arc BAD and BCD of the external circumference of the screw.
3. The form of the tooth profile developed must follow the basic rule of engagement *i.e.*, the common normal at the point of contact of both the profiles must pass through the point (polus). This will provide a uniform rotation. Correspondingly it will meet all the basic requirements.
4. The adjacent space between two teeth (*c*, *a*, *b* in figure 6.19) connecting each other must enclose themselves and forms separate chamber, for which cycloid profile the number of flow passages Z_1, in the driven is equal to that of driver with convex profile and a cycloid profile.

$$Z_1 = m(Z_2 - 1) \quad ...(6.29)$$

For Concave profile for the driver

$$Z_1 = m(Z_2 + 1) \quad ...(6.30)$$

Where m = No. of driven screws.

Z_2 = No. of flow passages in the driven screws.

Z_1 = No. of flow passages in the driver screw.

5. The length l of the casing enclosing the screw must be more than the axial length of the screws. $l > l_{min} = 1 - 4$ (Figures 6.9) *i.e.*, more distance between the two immediate next end of the space points 1 and 4. For the convex profile shape of the driver screw.

$$l > l_{min} = S_2 + \Delta l = t_2 + \left(\frac{\phi_2}{2\pi} - l_2\right) \quad ...(6.31)$$

6. The starting circle forming the screw with cycloidal profile must not pass through the surface ABCD, the two triangles, otherwise, profile will not be proper, and will be rotating with rubbing or abrasion between each other, which in due course change the profile configuration.

Failures to follow the above mentioned rules, lead to rubbing of screws, non-air tightness or a gap between spaces, which leads to the leak between delivery and suction chambers.

The most important part in screw pumps is the screw, its dimensions, and its profile configuration. The development of tooth profile of Mantelus screw is illustrated in figure 6.20. Circles I and II represent the circles of the driver and the driven.

In order to obtain a cycloid profile, the trajectory of point a_1, located on the circle II is traced, when it has a rolling contact with the circle I. When circle I rolls over circle II, point a_2 follows the epicycloid path a_2b_2, which is the tooth profile of the driven. In order to obtain the profile shape for the driver, circle II is rolled over circle I and the trajectory of point a_1 is drawn, epicycloid path a_1b_1.

The contour of the profiles a_1b_1 and a_2b_2, created by the trajectory of points a_1 and a_2 form two pair, two point profile, which provides leak proof, continuous contact passage for flow of liquid between two teeth of the driver and driven. When points a_1 and a_2 coincide each other, the points are designated as $a_1(a_2)$ or $a_2(a_1)$. The cycloid profile c_1 a_1 (a_2) c_2 forms the line of contact of the profiles. The curve a_1 (a_2) c_1 is obtained as the path, traced by the point a_2, on the driven screw, which also represents the clearance line, between screws. In the same manner the curve a_2 (a_1) c_2 is obtained as the path traced by the point a_2, on the driven screw, which also represents the clearance line. Thus two clearance lines are obtained between the driver and the driven. The edge point a_2 of the driven will be sharp, which when in contact with the driver screw surface creates higher friction losses, and the sharp edge becomes blunt and curved shape. The volumetric losses increase and the volumetric efficiency reduces. This point a_2 is corrected and, will be made into a curve.

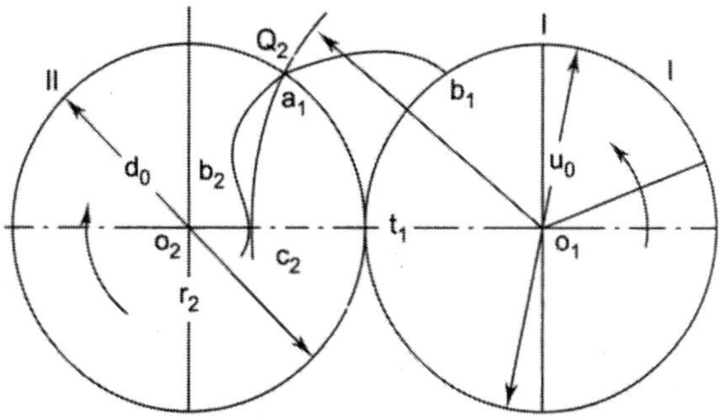

Figure 6.20. Construction of a epicycloid profile

The thickness of the driver screw at this point is increased, and thus corrected profiles are obtained. The dimensions and shape of the profile are determined in order to obtain proper clearance between the screws, and the driven screw will not have any resistive moment due to the liquid pressure acting on the screws. The mechanical losses between screws are at a minimum, and at the same time proper strength is obtained. Figure 6.7(*b*), is drawn based on the above procedure. Referring to figure 6.7, diameter 'd_2' is the reference diameter. Based on this diameter, the pitch and all other dimensions are determined.

The pitch $$t = \pi d_2 \tan = \beta = \frac{10}{3} d_2 \qquad \text{...(6 32)}$$

Where $\frac{10}{3} = \pi \tan \beta$ and β is the pressure angle of the screw.

6.13 PROFILE CONSTRUCTION

In order to obtain a continuous contact between the driver and the driven screws at the time of meshing and also to obtain a leak proof effect between suction and delivery chambers, the profile shape development, perpendicular to the axis, can be obtained only when the line of meshing point follows a curved path at the outer surface. The two contact curves BA_1D and DCB are obtained only with a symmetrical cycloid profile, which lies on one side of the starting point of the external surface (figure 6.21).

Profile contours for the screws 1 and 2 are obtained by moving the circle I-I over the circle II-II without any slip and similarly by moving the circle II-II over the circle I-I. When Point A_2, located on the circle I-I is rotated along the circle II-II, an extended epicycloid curve A_2E_2 is formed , which forms the tooth surface of the screw 2. When this developed curve, meets the starting circle I-I the point A_2 becomes point A_1, which is the angular contact point of the screw 1.

When circle II-II is rotated on I-I, the point E_2 on the circle II-II will give the development of shortened epicyclic curve A_1E_1. When this profile meets the starting circle II – II, point E_2 becomes the top surface point of the screw 2. The profile of the screw 1 and the passage of the screw 2 are symmetrical with respect to radius passing through the arc A_1A_1' and A_2A_2'. The length of the arcs A_1A_1' and A_2A_2' are related as

$$\frac{ArcA_1A_1'}{ArcA_2A_2'} = \frac{R_{e1}}{R_{i2}} \cdot \frac{r_2}{r_1}$$

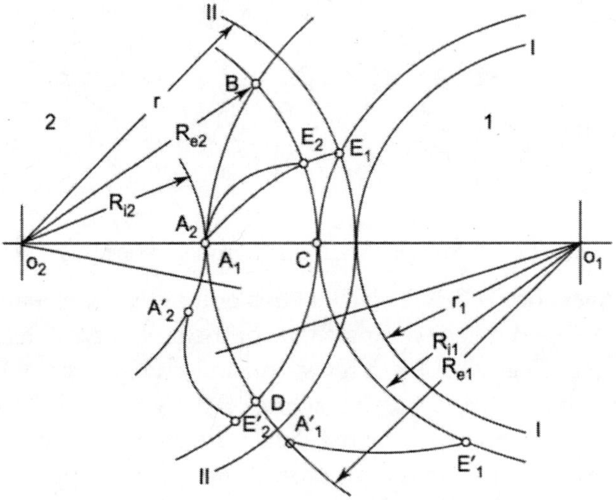

Figure 6.21. Construction of symmetrical epicycloid one sided profile (Sharp corners)

Radii r_1 and r_2 of the base circles of the screws are propositional to the number of teeth Z_1 and Z_2 respectively

$$\frac{r_1}{r_2} = \frac{Z_1}{Z_2}$$

The profile of the tooth for the screw 1 is drawn as per the shortened epicycloid A_1E_1 and $A_1^1E_1^1$ as well as the arc $A_1'A_1^1$ of the external circle R_{c1}. The passage in the screw 2 is drawn as per the extended epicycloids A_2E_2 and A_2E_2 and arc A_2A_2 of inner circle R_{c2}. The tooth of the screw 1 is located on the starting base circle I-I, and the tooth of the screw

2 is located on the starting base circle II-II. These two teeth are called symmetrical epicycloids single sided teeth. During rotation, the point E_2 of screw 2 will slide along the profile A_1E_1 of the screw 1 and at the same time Point A_1 of screw 1 will slide along the profile A_2E_2 of screw 2. Thus the line of contact becomes perfect and continuous. The line of contact of the profile A_2E_2 with point A_1 is the arc A_2B and point E_2 and the profile A_1E_1 and the point E_2 and the profile A_1E_1 will be arc CB. It is evident that the line of contact of the profiles having symmetrical cycloid one sided profile is an enclosed line passing through the arcs of the outer circle BCD and BA_1D_1 (Fig. 6.21). This profile theoretically ensures a positive sealed contact between meshing screws separating the suction and delivery chambers. Actually the sharp corner points E_2 and E_2' of the screw 2, due to constant and continuous running gets rounded off and produces a clearance between meshing gears, which increases leakage, reduces actual flow. Volumetric efficiency is considerably reduced. In order to avoid the reduction, in volumetric efficiency and actual flow, due to wear of the corner points E_2 and E_2' of the tooth, the corners are designed as round corners and the contact is made with the rounded profile of the tooth at the corners. The sharp edge of the profile of the tooth in the gear 2 is corrected. The profile of the passage of the screw 2 is modified as extended epicycloid A_2F_2 and $A_2'F_2'$ and hypocycloids F_2E_2 and $F_2'E_2'$ and the arc A_2A_2' of the inner circle R_{i2}. The profile of the tooth for gear 1 changes to shortened epicycloid A_1G_1 and $A_1'G_1'$ normal epicycloids. G_1H_1 and $G_1'H_1'$, shortened epicycloid H_1E_1 and $H_1'E_1'$ and arc A_1A_1', of the outer circle R_{e1}. (Figure 6.22)

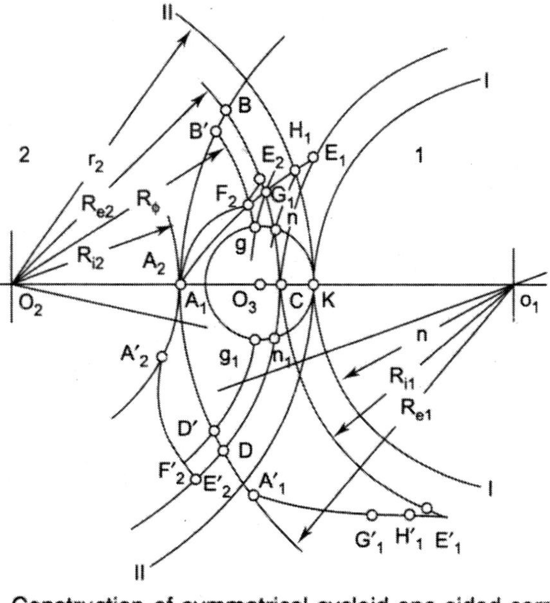

Figure 6.22. Construction of symmetrical cycloid one sided corrected profile

Hypocycloid F_2E_2 will form as the point of circle O_3, when it touches the circle II-II. Part of the profile G_1H_1 working together with F_2E_2 forms as a point of the circle O_3, when it touches the circle I-I.

The line of contact of the profile A_2E_2 and point A_1, is the arc A_2B_1' and point F_2 and profile A_1G_1 is the arc B'G, Profile F_2E_2 and G_1H_1 is the arc GH, point E_2 and profile H_1E_1

is the arc HC. The line of contact of the profile $A'_1E'_1$ of the second side of the tooth with profile $A'_2E'_2$ is the curve $A_1D'G_1H_1C$. The line of contact of the corrected screw is the enclosed line, but does not pass through the points B and D, the cross-section of outer circle of the tooth. That's why on the surface BB′ and DD′, the teeth passages of one screw is connected through the passage of the other screw. The delivery chamber of the pump is connected to the suction chamber, through this passage. The profile correction will not provide any perfect sealing between suction and delivery chamber. However, due to faster and considerable rate of metal removal at the sharp edges at the top of the profile of the screws, the clearance in uncorrected profiles will be larger than in the corrected profile.

While developing the profile with correction, it is essential that two angles should be formed by the curves BA_1DB (figure 6.23). Due to this teeth is formed on both sides of the starting base screws and hence called as symmetrical two sided cycloid profiles.

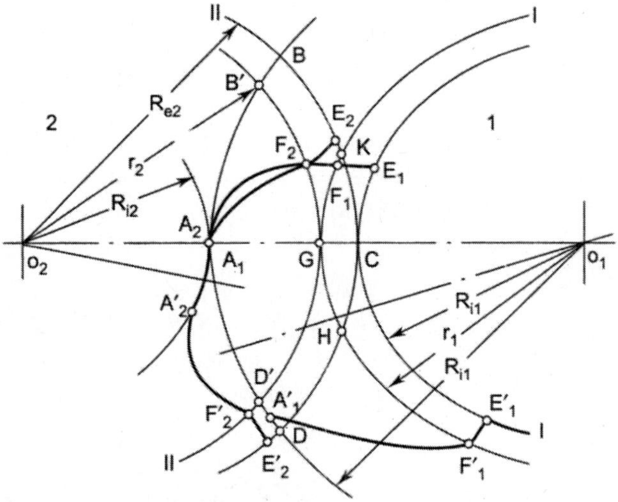

Figure 6.23. Construction of symmetrical cycloid two sided corrected profile

The side of the teeth of the screw 1 is formed by the epicycloids A_1F_1 and $A'_1F'_1$ and with the arc $A_1A'_1$ of the external circle R_{e1} and the profile nose is by the extended epicycloid F_1E_1 and $F'_1E'_1$. The profile of the passage of the screw 2 is developed by the extended epicycloids A_2F_2 and $A'_2F'_2$, by the epicycloids F_2E_2 and $F'_2E'_2$ and along the arc $A_2A'_2$ with inner radius R_{i2}.

The line of contact of point A_1 and profile A_2F_2 passes through the arc A_2B' and the point A_2 and the profile A_1F_1 passes through the arc B′G, point F_1 and profile F_2E_2 along the arc GH, point E_2 and profile along the arc CH. The line of contact of the profile $A'_1F'_1E'_1$ of the other side of the teeth with profile $A'_2F'_2E'_2$ passes through the arc $A_2D'GKC$.

Under equal distance between the centers of screw, the height of a two sided teeth will be more than that in single sided teeth. That's why; the application of two sided teeth provides an increase in area of the flow path between teeth and corresponding increase in flow rate, without increase in pump size. The only disadvantage in such two side screw is that the length of contact between teeth is considerably increased.

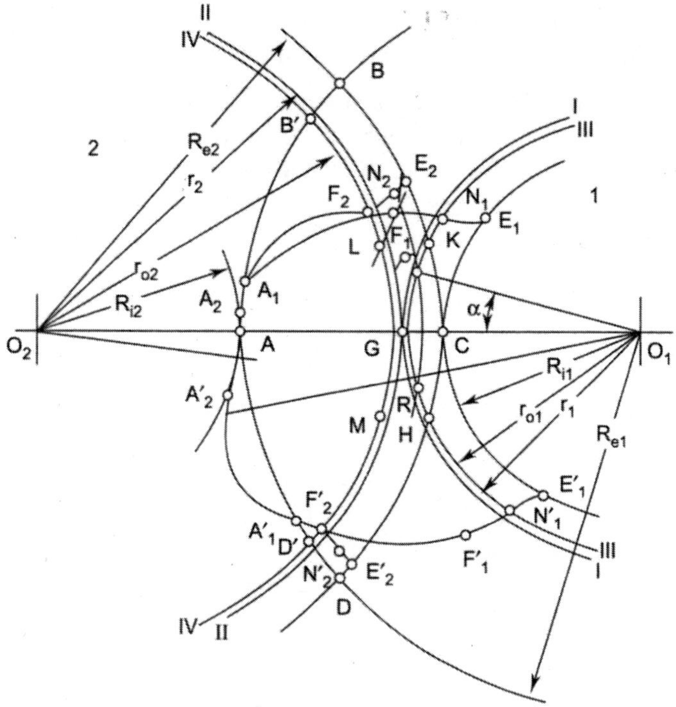

Figure 6.24. Profile formation with symmetrical two sided profile

In figure 6.24 the corrected profile development for the screw pump with two screws is illustrated. The profile of the tooth of the screw 1 is developed by the shortened epicycloid A_1F_1 and $A_1F'_1$ evolved with F_1N_1 and $F'_1N'_1$ of extended epicycloid N_1E_1 and $N'_1E'_1$ and along the arc $A_1A'_1$ with a radius of curvature R_{e1}. The profile passage formation of the screw 2 is developed by the shortened epicycloid E_2N_2 and $E'_2N'_2$ evolved with N_2F_2 and $N'_2F'_2$, the extended epicycloid F_2A_2 and $F'_2A'_2$ and arc $A_2A'_2$ with a radius of R_{i2}.

The line of contact of the point A_1 and profile A_2F_2 passes along the arc AB′, point F_2 and profile A_1F_1 along the arc B′L, profiles F_1N_1 and F_2N_2 along the straight line LR, the point N_1 and the profile N_2E_2 along the arc RH, the point E_2 and the profile N_1E_1 along the arc CH. The line of contact of Profile $A'_1E'_1$ on the other side of the teeth with the profile $A'_2E'_2$ passes along the line AD′MPKC

The dimensions of the screw illustrated in figure 6.24 are as follows:

The radius of the outer circle of the screw $R_{e1} = R_{e2} = R_e$.

The radius of cylinder of contact of driver screw $r_1 = 0.6\ R_e$ and driven screw $r_2 = 0.9\ R_e$.

The radius of curvature of the inner circle of the screw $R_{i1} = R_{i2} = 0.5\ R_e$.

The number of screws in the driver $Z_1 = 2$ with driven $Z_2 = 3$.

The path of the driver $s_1 = 3\ R_e$, driven $s_2 = 4.5\ R_e$.

The angle of contact of evolution of the screw =15°.

The area of the passage at the sideways A = 2.169 R_e.

In figure 6.25 the cross-section of the Profile for the teeth of the driver screw 1consists of shortened epicycloids a_1b_1 and $a'_1b'_1$, epicycloids b_1c_1 and $b'_1c'_1$ and the arc $a_1a'_1$ with diameter D_{e1}. the profile of the driven screw passage 2 consists of hypocycloid $c_2'b_2$ and $c_2'b_2'$, extended epicycloids b_2a_2 and $b_2'a_2'$ and the arc a_2a_2' with a diameter of D_{i2}.

The line of contact of the profiles c_1b_1 and c_2b_2 passes along the arc AE with centre O_3, point b_2 and profile b_1a_1 along the arc ED′, point a_1 and the profile b_2a_1 along the arc D′C. The line of contact of the profile $a_1'b_1'c_1'$ of the other side of the teeth with profile $a_2'b_2'c_2'$ passes along the arc B′C and B′FA.

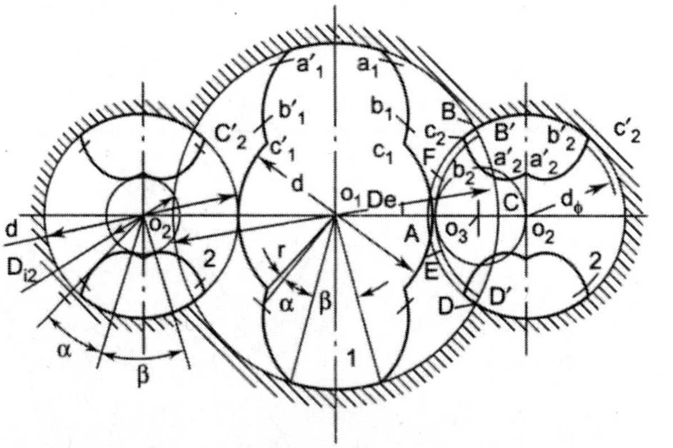

Figure 6.25. Formation of a three screw profile

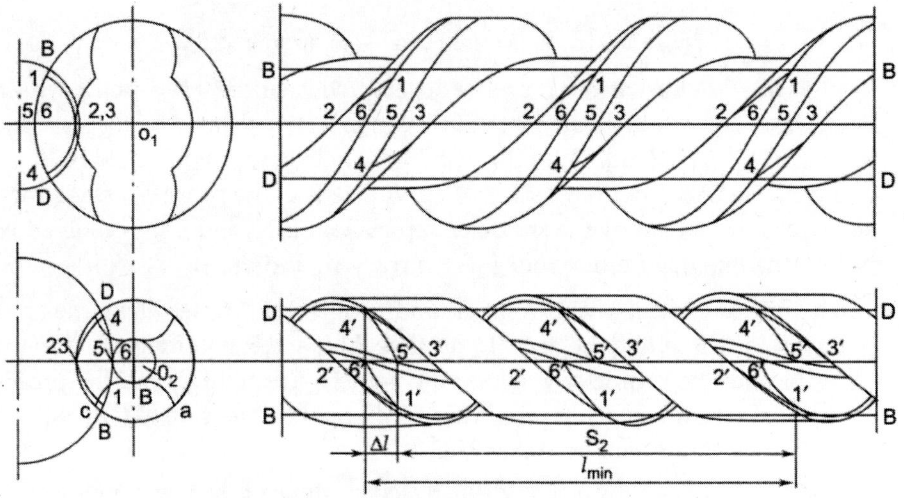

Figure 6.26. Line of contact between screws

The dimensions of the triple screw are:

Outer diameter of driver screw $D_{e1} = \dfrac{5}{3}\,d.$

Inner diameter of the driver screw $D_{i1} = d.$

Outer diameter of the driven screw $D_{e2} = d.$

Inner diameter of the driven screw $D_{i2} = \dfrac{1}{3}\,d.$

The starting diameter $\quad d_\phi = 0.95\,d.$

Angles, $\alpha = 0.4316, \beta = 0.18\pi, \gamma = 0.0536$

The length of the screw $S = (10/3)\,d$.

The area of the passage at the side $A = 1.2431d^2$

6.14 THE CONTACT LENGTH BETWEEN SCREWS

The driver and the driven screws of a triple screw pump is illustrated in figure 6.9. The contact surface is 12451 is shown at the side view and the related tangential contact along the linear length of the screw is shown 1234561. This contact line is repeated after the pitch length $t = \frac{s}{2}$. The screw edges 1′–2′ and 3′– 4′ of the driven screw tangentially touches the driver screw along the line 1–2 and 3–4.

The screw edges 6-1 and 4-5 of the driver tangentially meets the driven screw along the line 6′-1′ and 4′-5′. Along the line, 2-3(2^1-3^1) and 5-6(5^1-6^1), the screws tangentially meets with cylindrical surface of the passage and the outlet. The tangential line is a continuous line thereby provides perfect sealing.

The line BB and DD, illustrated in figure 6.9 are the edges of the casing under the screw. It is the boundary from which the screw comes out from the contact with the casing. Since the lines of tangential interaction of the corrected screw 1234561, do not end up to the line BB and DD on the surface 1-B(1′-B) and 4-D(4′-D) the passages will be adjacent to each other and axial sealing will not exist. However due to the smaller length of the side clearance, the leakage is very small.

6.15 DESIGN PROCEDURE FOR A SCREW PUMP

Pumps with two screws or three screws with one flow passage are quite common. The available parameter for the design of screw pump is flow rate (Q), Total Head (H) or total pressure (p) and the character of the pumping fluid such as viscosity, temperature etc., the suction pressure p_s is calculated from the suction pipe arrangement

$$p_s = p_{atm} - \gamma\left(H_s + \frac{C_s^2}{2g} + h_{fs}\right) \quad ...(6.33)$$

where p_s is the suction pressure, p_{at} is the atmospheric pressure, γ is the specific weight of the pumping fluid, C_s is the velocity flow in suction pipe, h_{fs} is the total minor and major friction losses in the suction pipe. Recommended axial velocity in suction C_s depends upon the viscosity, temperature and vacuum. The vacuum head, $H_s = \frac{p_{at} - p_s}{\gamma}$ and is selected from the standard graph figure 6.12. Normally the velocity C_s = 3 to 6 mps.

The area A is determined from the continuity equation $A = \frac{Q}{\eta_v \cdot C_s}$. Volumetric efficiency η_v lies between 0.7 and 0.95 the lower value is for high pressure units. Depending upon the flow and head, the type of profile and the number of units are selected.

From the area (A) the diameter (d) is determined as $A = m_A d$ where the coefficient m_A depends upon the profile type. For epicycloid profiles $m_A = 1.2431$. The preliminary diameter d for a three screw pump can be determined as $d = 2.44 \sqrt[3]{\frac{Q}{\eta_v n}}$

Outer radius of a two screw pump is determined as $R_e = 2.1 \sqrt[3]{\frac{Q}{\eta_v n}}$

The screws are selected as geometrically similar to suit the type. The speed n of the driver is selected from the axial velocity C_s and the passage s_1.

$$C_s = \frac{S_1 n_1}{60} \qquad \text{...(6.34)}$$

However the speed must agree with the prime mover speed. The working length of the screw determines the casing length and is determined based in the pressure head.

$$L = (1.5 \text{ to } 6)\, S_2 \qquad \text{...(6.35)}$$

For screw pumps, if the pressure $p > (4 \text{ to } 7)$ kgf/cm^2, an axial thrust reducing device will must be provided.

6.16 DETERMINATION OF THE SCREW LENGTH

The screw length and subsequently the length of the hub is determined by the pressure of the pumping fluid. Radial pressure increases if the pump delivery pressure is increased. This radial pressure acts on the hub. The unit pressure acting on the unit surface 'k_Z' should not exceed the mechanical strength of the screw material. By increasing the length (L) of the screw the unit pressure can be reduced. It is recommended that

$$L = \frac{1.53\, p d_H}{k_z} \qquad \text{...(6.36)}$$

And '$k_{y\delta}$' is gives as

^{0}E	3	12^0 and more
k_z	7	10kg/cm^2

At low pressure, L calculated as per the equation 6.15, is found to be lower than the pitch t. In order to provide continuous flow from suction to delivery chamber, the length 'L' is selected as $L = 1.25\, t$.

EXAMPLE

To determine the dimensions of three screw pump, given the flow rate $Q_a = 3$ Lps, pumping pressure, $p = 100$ atm, suction height, $H_s = 3$ mwc, viscosity of oil $^0E = 15$, volumetric efficiency = 0.8.

Referring to the graph (6.12), we get

$$C = 3.8\ m/sec \text{ (actual velocity at the tooth)}$$

$$C_a = \frac{3.8}{0.8} = 4.75\ \frac{m}{sec}$$

$$Q_{th} = \frac{3}{0.8} = 3.75\ Lps$$

Speed, $$n = 633.4\sqrt{\frac{4.75^3}{3.75}} = 3400 \text{ rpm}$$

The reference diameter (d_2) $d_2 = 244.05\,\frac{3.75}{3400} = 25.3$ mm

Other dimensions are

$$D_2 = \frac{5}{3} \times 25.3 = 42.3 \text{ mm}$$

$$D_1 = \frac{1}{3} \times 25.3 = 8.45 \text{ mm}$$

$$t = \frac{10}{3} \times 25.3 = 84.5 \text{ mm}$$

$$d_p = 0.67 \times 25.3 = 17 \text{ mm}$$

$$D_p = 1.82 \times 25.3 = 46 \text{ mm}$$

$$L = \frac{1.533 \times 100 \times 2.53}{10} = 39 \text{ cm}$$

CHAPTER 7

VORTEX PUMPS (REGENERATIVE PUMPS)

7.1 OPERATING RANGE

Vortex pumps are of two types

1. Pump impeller running inside a closed channel casing called centripetal pump.
2. Pump impeller running inside an open channel casing called regenerative pump.

These pumps are widely used, where discharge is relatively low, when compared to centrifugal pumps of same range, but higher than that of a positive displacement pump. These pumps possess low specific speed, n_s = 4 to 40. If a centrifugal pump is used in this region, it will be a multistage unit, with higher discharge.If a positive displacement pump is used, the discharge will be low. In both cases the efficiency will be low, power consumption will be high and the pump will be heavy. A slightly higher head with relatively low discharge is obtained from vortex pumps. Although, when $n_s < 40$, positive displacement pumps give higher efficiency than vortex pumps, economy in operation cannot be obtained. In vortex pumps, positive pumping effect is obtained. These pumps are used for sanitation, feed water treatment, condensation process industries etc.

7.2 PUMP WITH CLOSED CHANNEL IMPELLER—CENTRIPETAL PUMPS

Pump with the closed channel casing consists of an impeller a, mounted on a shaft, runs inside a closed channel of the casing *b* (Fig. 7.1(*a*)). In order to reduce the leakage, the axial clearance at the side, between impeller and casing, is kept very close (~ 0.06 mm). The impeller consists of a small cylindrical hub of constant diameter and axial width over which a number of lengthy radial blades of constant width and constant thickness are mounted. Number of blades vary from 12 to 24. On the side of the casing near the hub, inlet *c* and outlet *d* ports are provided. A side channel b is also provided at the casing, and near the outer edge of blades. It starts at the inlet port and ends at the delivery port. The channel is provided either on one side or on both sides of the impeller. At the central portion of the channel, the depth is larger, part of the circumference marked *c* in the figure 8.1(*d*), and constant for some circumferential length, whereas at the end, near inlet and outlet ports the depth is smaller, part of the circumference marked *a* for suction and *b* for the delivery, and equal to the clearance value between impeller and casing. The depth gradually reduces from the centre to ports. When the blades pass through the suction port, the liquid comes out of the inlet port, due to a suction effect created and occupies the impeller passage. This liquid is acted on by centrifugal force. It enters the side channel casing, due to the increase in depth of the channel, and also moves along the direction of rotation. Finally the liquid enters the delivery port due to the gradual reduction in the depth of the channel. In impeller channel, the liquid moves from hub to periphery at the inlet port and moves from periphery to the hub near the delivery port and leave through the delivery opening. These pumps are self priming, if it is already filled with liquid before starting the pumps. A rotating water ring is created in the impeller passage, which pushes the air and liquid out of the pumping passage area and from the suction pipe. It is essential that inlet mouth is always kept above the pump casing level, in order to ensure that water is always available inside the pumps.

7.2.1 Construction

In figure (7.1) a horizontal centripetal pump is shown. It consists of two casings namely suction and delivery, *b* in separate sections and connected by a through bolt from end to end. The impeller a is located at the space in between adjacent intermediate casings, with an axial running clearance of 0.06 to 0.1 mm between casing and impeller. The impellers are mounted on the shaft which is supported by the bearings either one at each end or

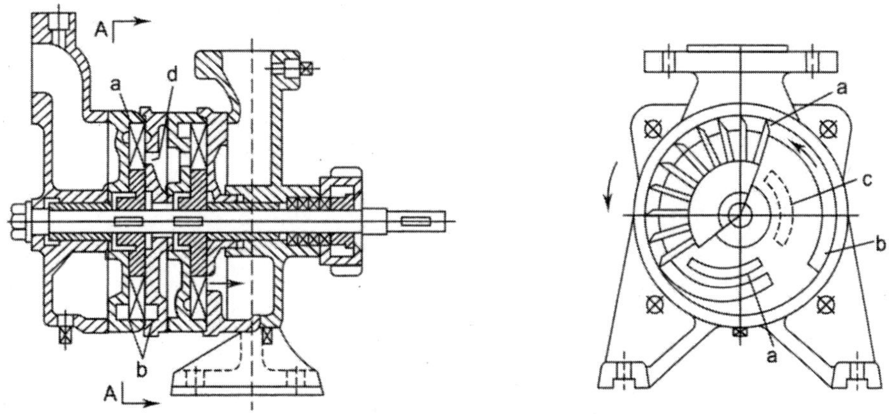

(a) Cantilever type

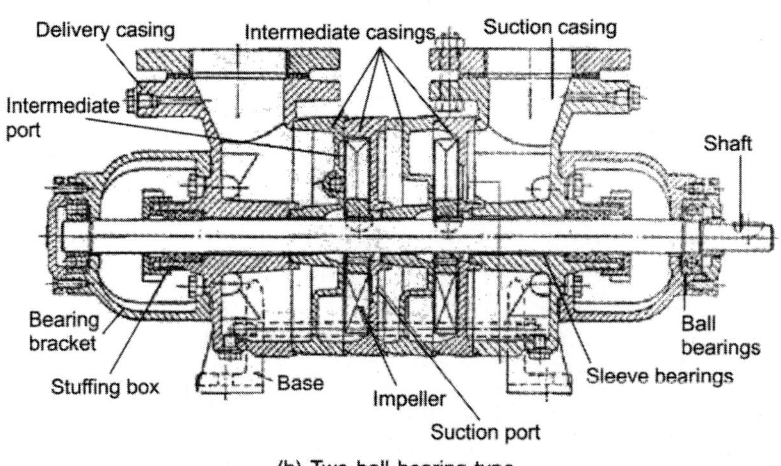

(b) Two ball bearing type

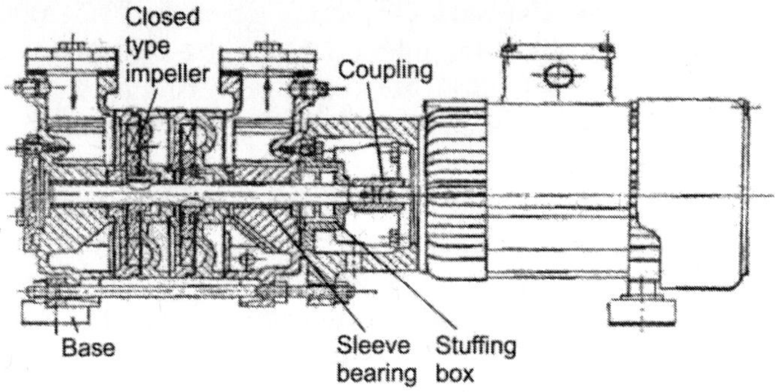

(c) Monoblock type

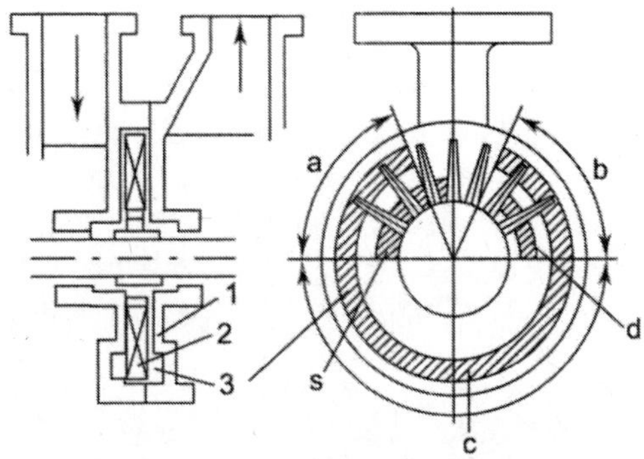

(d) Details

1. Casing. 2 .Impeller. 3. Channel opening in casing (a) Suction port (b) Delivery port (c) Intermediate channel (d) Delivery port

Figure 7.1. Closed Type Vortex Pump–Centripetal Pump

only at one end and with a sleeve bearing at the other end, (figure 7.1). In the later design, the intermediate casings, the suction and the delivery casings, impeller with shaft attached to the bearing bracket as overhung unit or cantilever unit. One end of the shaft is connected to the prime mover through coupling. The impeller is mounted on the shaft as free unit by which it can axially move freely, while running. A ball bearing support near coupling and two bush bearing supports at each end as well as at intermediate casings are provided to support; the shaft. In order to avoid the leakage of water out of the casing, stuffing box is provided with asbestos packing either at both ends or at coupling end only depending upon the construction adopted.

In all horizontal type constructions the suction and delivery mouths are kept well above the pump so that the liquid always remains in the pump when the pump is not working, which makes the pump a self priming unit. In most cases, these pumps are manufactured as multi stage units (2 to 3 stages). These pumps do not run at the high speeds and hence possess relatively low head, but higher than head developed by a centrifugal pump, for the same specific speed.

7.3 PUMP WITH OPEN CHANNEL CASING—REGENERATIVE PUMPS

The open channel casing type pump, called Regenerative pump, consists of a disc, on the periphery of which, a number of short radial blades are located. In some of the designs, inclined blades, instead of radial blades are also provided with the angle of 60° to 90° with respect to the central wall of the impeller. The suction and delivery ports are located radially at the casing and are separated by a circumferential wall in the casing.

Very close tolerance is maintained at the clearance, 0.05 to 0.2 mm, between the casing and the impeller, so that high pressure liquid from the delivery will not pass to the suction side chamber through this clearance when the pump is running. Number of vanes is selected usually vary from 24 to 32. The radial clearance is 0.06 to 0.2 mm and the axial clearance is 0.05 to 0.1 mm between casing and impeller.

The liquid flowing through the impeller channel and through the casing channel are at different angles as well as at velocities with respect to the axis of the shaft. A part of the liquid located in the impeller, always rotates at angular velocity of ω_i which is equal to the impeller angular velocity, and a part of the liquid at the side channel in casing moves with an angular velocity, ω_c. The difference between these two centrifugal forces creates a circulating, vortex flow in the meridinal direction with a velocity of C_m. This velocity is radially outward at impeller channel, moves to the outer periphery area in the impeller and then flows down to the impeller suction of the next blade. The kinetic energy is converted into potential energy in the casing channel. Likewise, the energy is increased and converted into potential energy when liquid passes through each blade system.

The impeller blade system (9) is separated by a central wall (5) as left side and right side. On the periphery of the impeller (2) and on both sides of the wall a number of radial blades (6) are provided. Number of blades vary from 24 to 32. These types of impellers are called regenerative impellers. These are different from impeller in a closed type casing, where there is no intermediate separating wall (5). The impeller is mounted on the shaft (8) and runs inside a casing (2) which contains suction and delivery mountings as well as flow channels (4). A closed tolerance at the clearance between the impeller and casing at the side (7) is maintained, 0.05 to 0.2 mm, in order to make the leakage reduced to a minimum, due to high pressure difference. The impeller is enclosed by the casing at the periphery, where the impeller blades are exposed free inside the channel (4) of the casing (2). The casing channel (4) is connected to the suction and delivery (1) pipe at radius r_3 through respected opening of the casing mouth (1). The delivery and the suction openings at the casing are separated by a radial clearance (10) of $r_3 - r_2 \approx 0.15$ to 0.2 mm, between the casing and impeller.

When the pump is filled with water and runs, the passage (figure 7.3(*b*)) between two blades of the impeller, also rotates. Due to centrifugal force the liquid simultaneously moves towards the periphery, as similar to the flow in centrifugal pump. A part of the liquid from the inlet port (12) enters the (inlet end) lower part of the blade and gradually moves up along the passage, relative to the blade. Referring to the figure 7.3(*b*), the line A-M *i.e.*, the flow from the inlet port (12) to the inlet edge of the impeller (located at the lower radius) is an absolute flow of liquid. Further flow takes place inside the blade passage (4) which is a relative flow. Fluid gets the energy from the blade moment of momentum. The flow is given the line m-B-c. After passing through the blade system the fluid circulates inside the casing channel and then enters the next vane passage. This flow is indicated by the line C-M.

Likewise, the liquid passes through the impeller passage channel a number of times. Each time the fluid gests additional energy there by the total head is increased each time for the same quantity of flow. This is similar to that in a multistage centrifugal pump. The flow in regenerating pump is a vortex flow *i.e.*, a screw type circulating flow with and trajectory ABCDEF. These open type impellers can take a very high suction of 3 to 3.5 MWC. However it depends upon the circumference length of the radial clearance between delivery mouth and suction port or mouth.

The main advantage of these pumps when compared to the centrifugal pumps is that more impeller passages are provided in one and the same impeller and more head is obtain for the same speed. Total weight of the pump is considerably reduced by 3 to 5 times. However the flow rate is very small. Regenerative pumps are more economical, and are used for very high head and for low flow rate, where reciprocating pumps and centrifugal pump cannot be used.

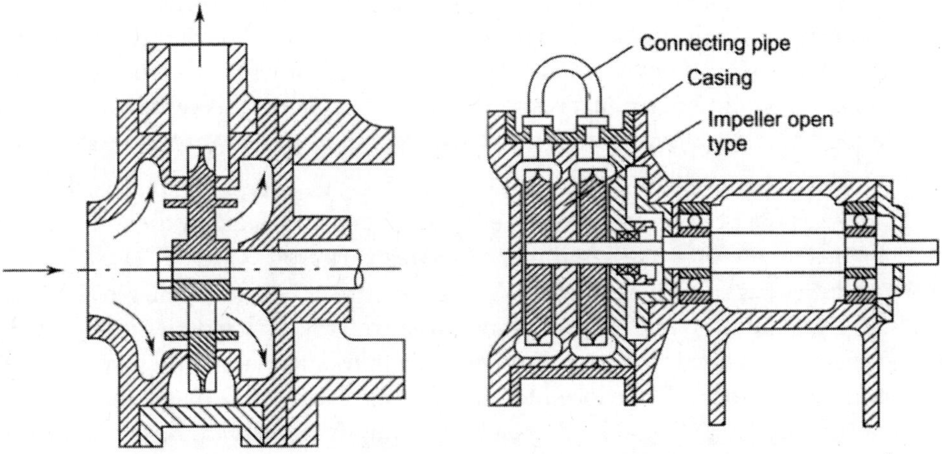

(a) Single stage (b) Two stage

One end supported type (cantilever type)

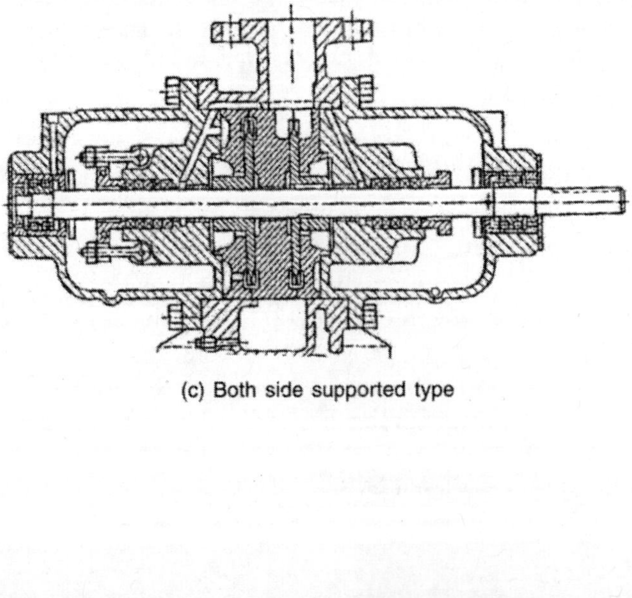

(c) Both side supported type

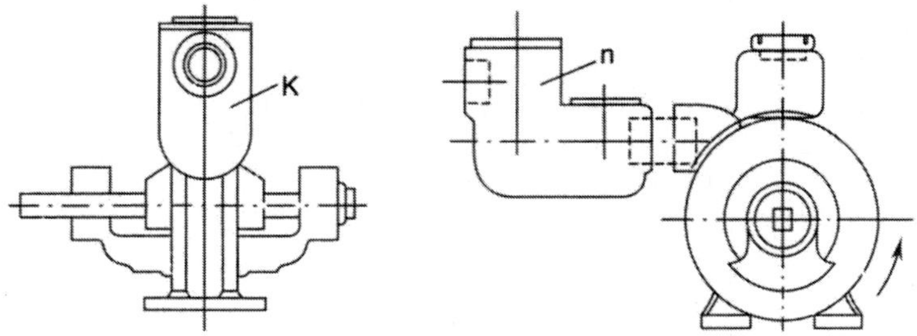

(d) Self priming attachment for regenerative pump

Figure 7.2. Open Type Vortex Pump – Regenerative pump

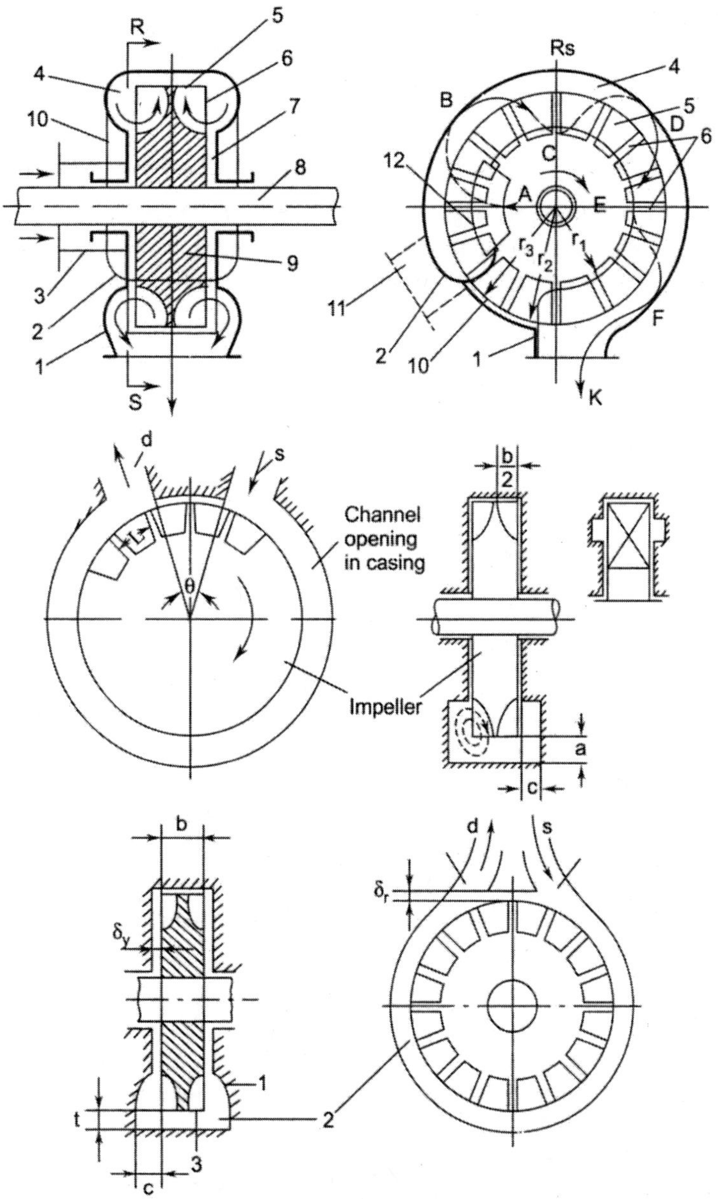

Figure 7.3. (a) Schematic diagram of a vortex pump

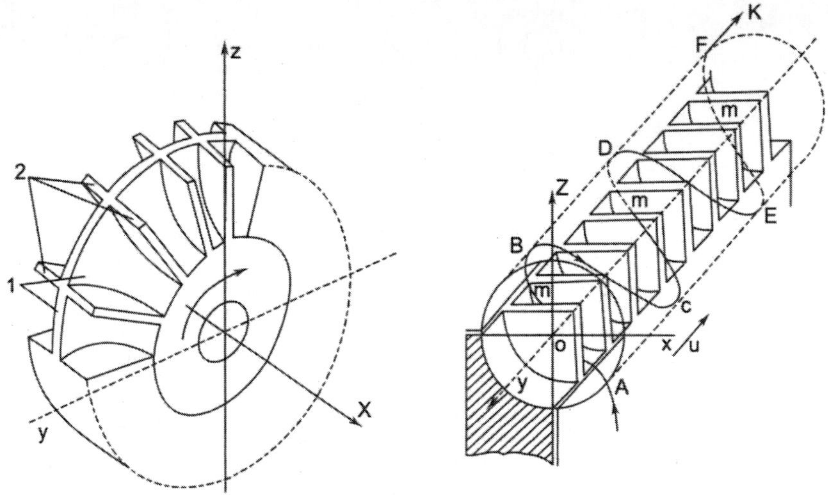

Figure 7.3. (b) Fluid flow through impeller blades of a regenerative pump

The construction of these pumps is very simple. The efficiency of the pump varies from 40 to 50%. These pumps occupy the specific speed n_s = 4 to 50. These pumps are used in combination with centrifugal pump there by n_s and η are improved.

Regenerative pump construction is simple when compared to that of centrifugal pump. Due to positive displacement effect, these pumps have very good suction characteristics. These pumps are used as priming pump and connected to the centrifugal pump in series. In figure 7.2 (*a* and *b*), a single stage and a two stage regenerative pump are illustrated. The suction and delivery mouths can be accommodated either in radial direction or in axial direction as the case may be.

The regenerative pump is very simple, less weight per litre of pumping fluid. The manufacturing process, of these pumps, is very simple. These pumps are very economical in weight, for the range of discharge and head selected. No other pump can match this area of head and discharge. Each channel in this pump acts as an individual impeller. One regenerative impeller acts as many small impellers in a multistage centrifugal pump.

These pumps are used for pure water or for low viscosity fluid like kerosene, petrol etc. These pumps run at very high speeds of about 6000 rpm. The main advantage of these pumps are higher head, approximately 3 to 6 times that of a centrifugal pump, running under similar conditions. The pump weighs considerably less when compared to centrifugal pump. The disadvantage is the poor efficiency and smaller suction head of operation. The radial force of both pumps is very high when compared to that of a centrifugal pump. Regenerative pump is a non positive unit, hence requires additional priming equipment, low cavitation characteristics and hence should be run at a lower speed.

One of the drawbacks in centrifugal pump is poor suction characteristics. Pump cannot suck water from the suction sump even though the pump is filled with water. When starting the pump, both pump and the suction pipe must be completely filled with water. In order to remove the air in the suction pipe, sometimes a positive displacement pump is fitted before the centrifugal pump. The overall efficiency of the system is reduced. Regenerative pumps are adopted in such conditions to match. These pumps are normally used for pumping clear water. If water or liquid with solids in suspension are pumped, the running clearance between the impeller and casing becomes larger due to gradual abrasion and pump stops. In such cases, the pump material should be selected to withstand abrasion.

Regenerative pumps are used for flow rate upto 100 m^3/hr. Economical range is approximately 35 m^3/hr. These pumps are uneconomical for discharges above 35 m^3/hr. Centrifugal pumps are better than regenerative pumps in these ranges. Minimum flow rate can be 0.4 m^3/hr. Pumps are made even upto 150 lit/hr. Economical Head is approximately 200 m. However pumps are manufactured even for 500 m total head.

7.3.1 Principles of Operation

The principle of operation of regenerative or vortex pump is the cumulative increase in the energy impulse of the liquid, when the liquid flows through each channel of the impeller by recirculation of liquid through the hollow passage in the casing, The total head increases due to the impulse given by the impeller at each passage, when liquid passes from bottom side of the flow passage, which is called the suction side, to the topside of the flow passage, which is called the delivery side of the passage. The liquid, then circulates inside the casing passage, in a radially inward direction and is brought to the suction side of the next flow passage. Thus the total head of the liquid is increased at each passage. If there are 'Z' number of passages (which is equal to the number of blades) and (h_o) is the head developed at each passage the total head H is H = (Z × h_o).

7.3.2 The Importance of Pump

Due to the provision of a number of passages in impeller, the total head is higher, nearly 4 to 5 times in regenerative pump than that in centrifugal pump for the same speed of the pump. This reduces the size of the pump considerably in comparison with centrifugal pump for the same head. Regenerative pumps have good suction characteristics. The performance characteristics of regenerative pump of 100 mm diameter and a centrifugal pump of 150 mm diameter, tested at the same speed of 1750 rpm is given, in figure 7.3. It is evident from the graph that the regenerative pump gives higher efficiency almost 2 times than centrifugal pump for the same discharge of 5 m^3/hour. The H-Q curve of a regenerative pump has a steep drooping characteristics, where as, centrifugal pump has a less drooping H-Q characteristics, (figure 7.4). Viscosity of the pumping liquid affects the performance of the regenerative pump if the viscosity is more than 15°E. The head drops more and the power increases higher. That is why regenerative pumps are not recommended for high viscous liquid pumping.

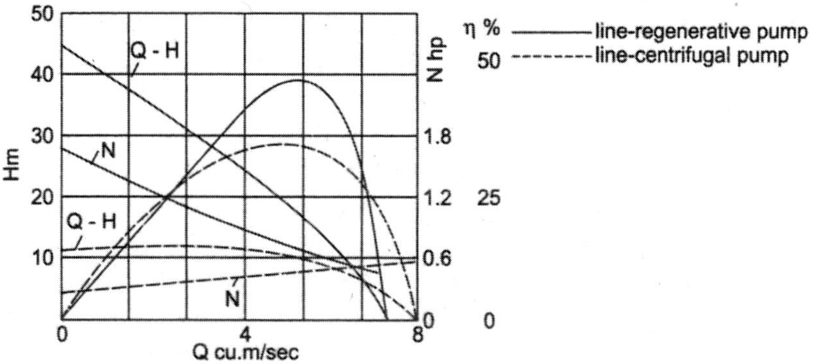

Figure 7.4. A comparison between centrifugal and regenerative pump characteristics

7.4 CALCULATION OF TOTAL HEAD

The theory of regenerative pump is the same as that of the theory of roto dynamic pumps. Euler's equation is used to determine the total head. From the experiment analysis it is found, that the total head increases, with the number of blades and the length of flow passage in impeller (4). The flow takes place in regenerative pump in a complicated manner. There is (1) a vortex flow (rotary) relative to the axis of figure 7.3 and (2) vortex flow relative to pump axis. Both these flows combined together give a trajectory ABCDERK (figure 7.3(*a*) and (*b*)). In casing channel, the flow is in the form of a screw, which is responsible for the energy increase by way of impulse. Apart from that part of the liquid also undergoes vortex flow relative to OX(figure 7.3) in the impeller passage (relative vortex). This additional vortex is added with the main vortex (relative to OY). Both these put together increases the total head.

$$H_{th} = \frac{C_{u2}\,u_2}{g}\,(\text{for line A}b) + \frac{C_{u2}\,u_2}{g}\,(\text{for line CD}) + \frac{C_{u2}\,u_2}{g}\,(\text{for line EF}) = 3\frac{C_{u2}\,u_2}{g}.$$

$$\text{In general} = i\,\frac{C_{u2}\,u_2}{g} \qquad ...(7.1)$$

where '*i*' is the increase due to the flow in impeller passage in succession. Due to finite number of blades, the head reduction is accounted by a coefficient 'μ'

$$H_{th} = i\mu\,\frac{C_{u2}\,u_2}{g}$$

Since

$$H = H_{th}\,.\,\eta_h = i\mu\eta_h\,\frac{C_{u2}\,u_2}{g} \qquad ...(7.2)$$

Taking

$$C_{u2} = \frac{\lambda}{2}\,u_2 \quad H = \Psi\,.\,\frac{u_2^2}{2g} \text{ where } \Psi = i\mu\eta_n\lambda, \qquad ...(7.3)$$

For regenerative impellers, the value, $\Psi = i\mu\eta_n\lambda$, depends upon the number of blades, form (dimensions) of the channels, both in impeller and in casing especially the channel width (in figure 7.3) and roughness of the surface.. Theoretically, an equation, as similar to centrifugal pump can be developed for vortex pump. Because of absence of the correct value of 'μ', the H-Q curve is always obtained by experiment. H-Q curve for the regenerative pump is steeper than that of centrifugal pump. This is because of more circulating motion in between casing and impeller channel, at reduced flow rate. A breaking effect created inside the pump passage due to throttling of the control valve at the delivery for flow reduction. The co-efficient '*i*' is increased, which in turn increases the total head. In the figure 7.5, curves 1 and 2 are for different speeds $n_1 > n_2$. Curves 2 and 3 are for different number of blades, $Z_2 > Z_3$. Curve 4 is for power and curve 5 is for efficiency for impeller number 1./1, 17/

From the curve, it is seen that, number of blades influence on total head of the pump. The energy increase takes place at every stage when fluid passes through the blade passage, by recirculation. This pressure increase is proportional to the angle of movement, in the circumferential direction φ, (figure 7.6(*b*)). Taking moment for the elementary length '*dl*'. If the area is 'A_1'at inlet and 'A_2' at outlet of the impeller and the two side areas are designated as A_3 and $A_4 (A_3 = A_4 = A_s.ch)$ and the side area is A_5 (fig. 7.6(*b*)) and considering the ideal, steady and symmetrical liquid flow, relative to the axis of impeller and taking the side channel as linear,under steady flow through side channel, the sum of all surface forces acting in the elementary mass is equal to the change of moment of momentum of the mass flow of the fluid per second passing through the surface A. Containing the defined mass

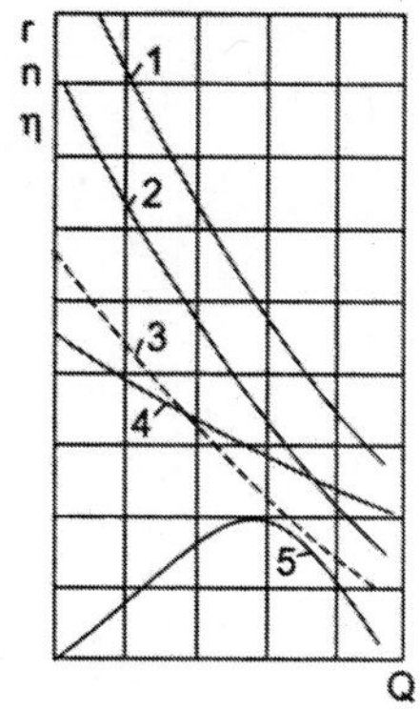

Figure 7.5. Effect of number of blades and speed on the performance of the impeller

$$pA_3 - (p + dp)\, A_3 = \frac{\gamma}{g} r.dl.\, C_m . C_{u1} - \frac{\gamma}{g} r.dl.C_m . C_{u2}$$

$$dp = \frac{\gamma\, r dl C_m}{g A_3} \left(C_{u2} - C_{u2}\right) \qquad ...(7.4)$$

Where, p and $(p + dp)$ are the hydraulic pressures on the side channels A_3 and A_4, Cm is meridinal velocity of the fluid and C*u* is the tangential velocity of the fluid at outlet A_2 and at inlet A_1 respectively

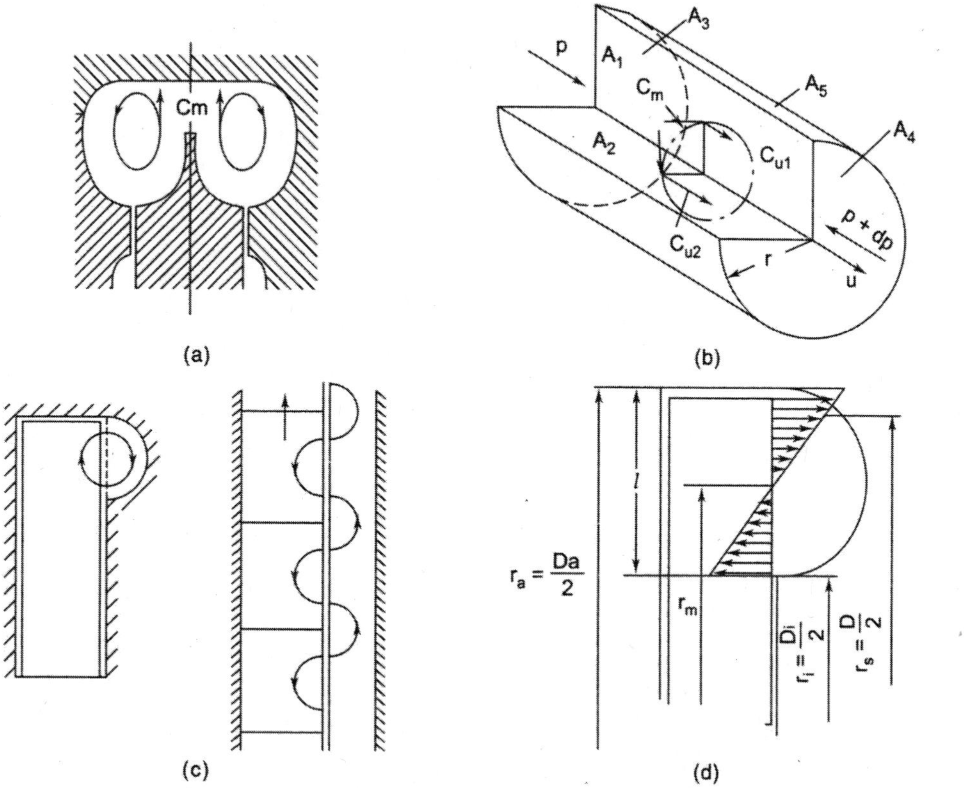

Figure 7.6. Schematic diagram for the determination of the total head

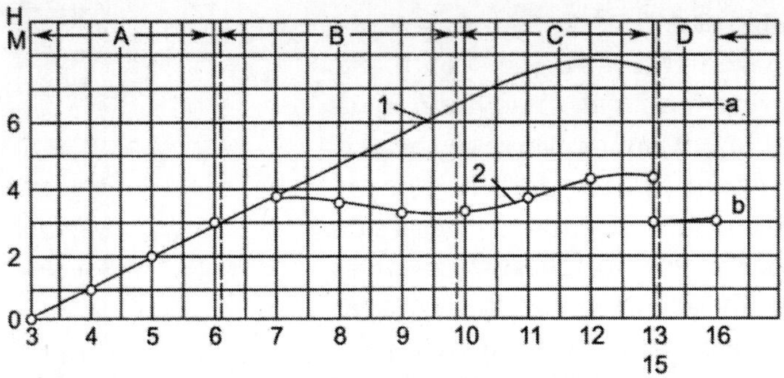

Figure 7.7. (a) Pressure distribution along the side channel

1. Continuous line is for Centripetal pump (closed type impeller), (2) Dotted line is for the regenerative pump (open type impeller). A. Suction port, entry to the blade inlet, C. Delivery, exit from t he blade B. Connecting (intermediate) passage, and D. Delivery port. Refer figure 7.1

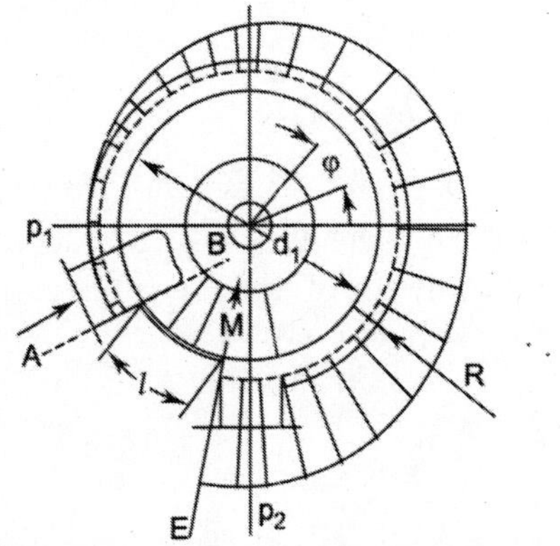

Figure 7.7. (b) Increase in pressure along the circumference of the impeller of the regenerative pump

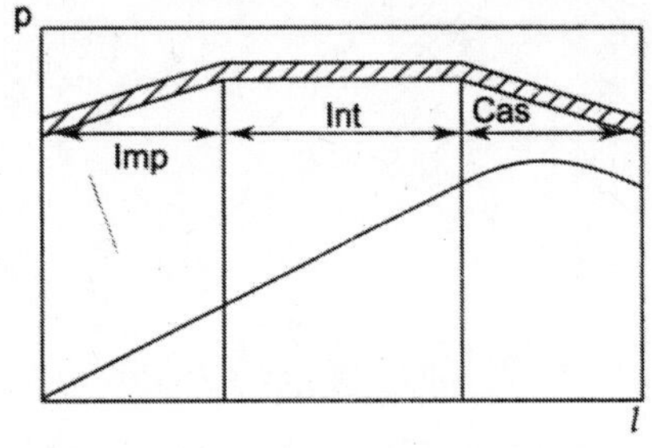

Figure 7.7. (c) in both passages

In figure 7.7(*b*), the flow velocity vector diagram, in figure 7.7(*a*), (*b*) and (*c*) pressure distribution in both passages, in figure 7.7(*a*) and (*c*) path of the fluid travel, through impeller passages are illustrated.

The surface force is the hydrodynamic pressure '*p*'. In equation (7.4) only surface forces are acting on the areas A_1, A_2 and A_5, and directed normal to the tangential direction axis. The moment of flow in channel and resolved components of these forces will be zero both in magnitude and direction.

The liquid under consideration flows through area A_2 and A_3. The flow rate and $Q_{in} = C_m A_2$ and $Q_{in} = C_{u1} A_3$ is the flow rate at outlet through the area A_1+ and A_4. Mass flow rate $m = \frac{\gamma Q}{g}$. Due to symmetry $\frac{\gamma Q}{g}(C_{u1})_{A4} - \frac{\gamma Q}{g}(C_{u1})_{A3} = 0$

or $(C_{u1})_{A4} = (C_{u1})_{A3}$ and

$$F_{in} = \frac{\gamma Q_i}{g A_3}(C_{u2} - C_{u1}) \quad ...(7.5)$$

The pressure at the side channel increases in proportion to the length of the channel. Total pressure increase

$$\Delta p = \frac{\gamma' L C_m}{g A_3}(C_{u2} - C_{u1})$$

Theoretical head developed by the regenerative pump will be

$$H_{th} = \frac{\Delta p}{\gamma} = \frac{\gamma L C_m}{g A_3}(C_{u2} - C_{u1}) = \frac{Q_{in}}{g A_3}\left(C_{u2} - \frac{Q}{A_3}\right) \quad ...(7.6)$$

Actual head H = H_{th}-losses. The total loss will be the loss in the side channel. The quantity $Q_{in} = C_m rl$ is the quantity entering from the impeller to side channel and then after circulation increasing channel enters the quantity which undergoes the vortex flow in meridianal section useful power output

$$N_{oth} = \gamma Q H_{th} = \frac{\gamma Q_{in}}{g A_3}\left(C_{u2} - \frac{Q}{A_3}\right) \quad ...(7.7)$$

Power, given to the fluid by the vortex motion inside channel in casing, during the process

$$N_i = \frac{\gamma Q_{in}}{2g}\left(C_{u2}^2 - C_{u1}^2\right) = \frac{\gamma Q_{in}}{2g}\left(C_{u2}^2 - \frac{Q^2}{A_3^2}\right) \quad ...(7.8)$$

Considering losses in the casing due to mixing and the losses in the side channel

$$\eta_i = \frac{N_0}{N_{in}} = \frac{\gamma Q H}{\frac{\gamma Q_p}{2g}\left(C_{u2}^2 - \left(\frac{Q}{A_3}\right)^2\right)} = \frac{2}{\frac{C_{u2}}{C_{u1}} - 1}\frac{H}{H_{th}} \quad ...(7.9)$$

Total losses will be the sum of the losses in the flow passage in the N_{hp}' losses with leakage inside the pump 'N_e" and the chemical losses 'N_m"

$$\eta_{imp} = \eta_o \eta_h \eta_m = \frac{N - (N_{LP} + N_l + N_m)}{N} \quad ...(7.10)$$

Overall efficiency of the pump $\eta_{imp} = \eta_{imp}\eta_{in}$

N.N Kupriuashin gave a relation between 'C_m' and the geometrical characteristics of the pump as

$$C_m = \frac{m A_{in} \omega_{in}}{\sqrt{Sa}}\left(1 + \frac{Q}{A_3 C_{u2}}\right)\sqrt{1 - \frac{Q}{A_3 C_{u2}}} \quad ...(7.11)$$

$$H_{th} = \frac{\Psi L}{gr} \cdot \frac{r_u}{R} \frac{mA_{im}}{A_3} \sqrt{\frac{S}{a}} C_{u2}^2 \left[1 - \left(\frac{r_u}{R} \cdot \frac{Q}{A_3 \cdot C_{u2}}\right)\right] \times \sqrt{1 - \frac{Q}{A_3 \cdot C_{u2}}} \quad ...(7.12)$$

$$\eta_i = \frac{N_0}{N_{in}} = \frac{\gamma QH}{\gamma HA_3 \, R\omega_{imp}} = \frac{Q}{A_3 \, \omega_{imp} \, R} \quad ...(7.13)$$

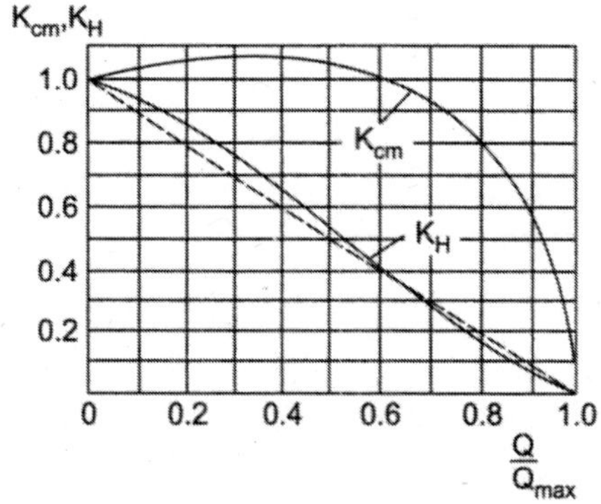

Figure 7.8. Graphs

$$k_{CT} = \frac{C_m}{(C_m)_{Q=0}} = h\left(\frac{Q}{Q_{max}}\right)$$

$$k_H = \frac{H'}{H'_{max}} = f\left(\frac{Q}{Q_{max}}\right)$$

Where m is the co-efficient for the channel ($m = 2/\pi$ for a circular channel), S is length of vane, Ψ is area reduction co-efficient due to blade thickness, r_u and R is the gravity centre of the area A_{imp} and the side channel A_3 (or A_4) respectively. Considering equation (7.11), the vortex velocity or the circulation velocity 'C_m' depends upon the shape of the side channel, number of blades and rate of flow, but independent of the pressure inside the side channel. C_m slightly increases from Q = 0 to 0.6 Q_{max} and then steeply falls when Q = 0.6 to 1.0 Q_{max} (figure 7.8).

It is found that the total head H has a direct proportionality, *i.e.*, linear with the discharge. It can also be concluded that C_m is independent of Q.

7.5 THEORY OF REGENERATIVE PUMP

The change of momentum in regenerative pump takes place in an accelerated manner, due to rapid mixing of liquid when it passes through casing channel, and a slow mixing when it is flows through the impeller passage. In regenerative pumps two types of flow exist: (1) circulatory flow, due to the difference in pressure between the impeller passage and the channel in casing and (2) Main flow, due to the pressure difference between working (leading) and non working (trailing) surfaces of the blade.

Considering the interchange of moment between the impeller and the fluid, as similar to the centrifugal pump, the external moment acting on the impeller is equal to the difference between the moment of the fluid between inlet and outlet of the impeller.

$$M = M_2 - M_1 = \frac{\gamma}{g} Q_r r (C_{u2} - C_{u1}) = \frac{\gamma}{g} 2\pi r b \alpha C_r r (C_{u2} - C_{u1}) \quad ...(7.14)$$

where $\frac{\gamma}{g} Q_r = \frac{\gamma}{g}$ **$2\pi r\, b\alpha C_r$** is the mass of fluid flowing radially through the passage.

r is the average radius of the blade and the side channel.

$\alpha = \frac{360 - \theta}{360}$ is the active portion of the channel.

C_r is the radial velocity of flow in the channel. For Regenerative pump $C_r = C_m$ C_{u2} and C_{u1} are the average tangential velocity component of the absolute velocity at outlet and at inlet respectively.

Moment **M** created by the Hydraulic force acting on the casing is equal to

$$M = (p_2 - p_1)\, Ar \quad ...(7.15)$$

Where p_2 and p_1 are the pressures at the beginning and at the end of the casing channel and, A is the area of cross section of the channel.

Equating the above two equations 7.14 and 7.15

$$\frac{\gamma}{g} Q_r r \left(C_{u2} - C_{u1}\right) = \left(p_2 - p_1\right) Ar \quad ...(7.16)$$

The theoretical head Hth of the pump is

$$H_{th} = \frac{p_2 - p_1}{\gamma} = \frac{Q_r}{gA}\left(C_{u2} - C_{u1}\right) = \frac{Q_r}{gA}\left(C_{u2} - C\right) \quad ...(7.17)$$

The flow quantity **Q**, correspondingly the velocity **C = Q/A**, through the impeller remains constant at all flow rates **Q**, across the side channel, since it doesnot depend on the pressure at the side channel and remains same at inlet and outlet of the impeller also. The velocity '**C**' remains constant at all regions of operation. Equation (7.15) gives the relation between total Head H_{th} and the flow rate **Q** which is a straight line, drooping down, starting from Y-axis,where **Q = 0** to X-axis, where H = 0. Maximum quantity $\mathbf{Q_{max}}$ will be when **H = 0**. Applying this to equation (8.15).

$$0 = \frac{Q_r}{gA}\left(C_{u2} - \frac{Q_{max}}{A}\right), \quad ...(7.18)$$

from which

$$C_{u2} = \frac{Q_{max}}{A} \quad ...(7.19)$$

Maximun Head $\mathbf{H_{max}}$ will be when Q = 0 and C = 0. Substituting this value in the equation 7.4

$$H_{max} = \frac{Q_r}{gA}\frac{Q_{max}}{A} = \frac{Q_r\, Q_{max}}{gA^2},$$

from which

$$Q_r = \frac{H_{max}\, gA^2}{Q_{max}} \quad ...(7.20)$$

The outlet angle $\beta_2 = 90°$, since the blade is radial at the outlet. $C_{u2} = 0.85\ u$. The value C_{u2} is less than u, indicates the influence of the finite number of blades. For blades having β_2 other than 90°, the value C_{u2} is determined.

$$C_{u2} = u \pm \frac{C_r}{\tan \beta_2}$$

As per the experimental results, the velocity C_{u2} is approximately equal to the effective peripheral velocity with radius r_s (figure 7.9)./ 18,25 / Usually the value $C_{u2} = 0.85\ u$ for radial blades, in order to take into account the effect of finite number of blades. The radius $r_s = r_m + ((r_a = r_i)/3)$

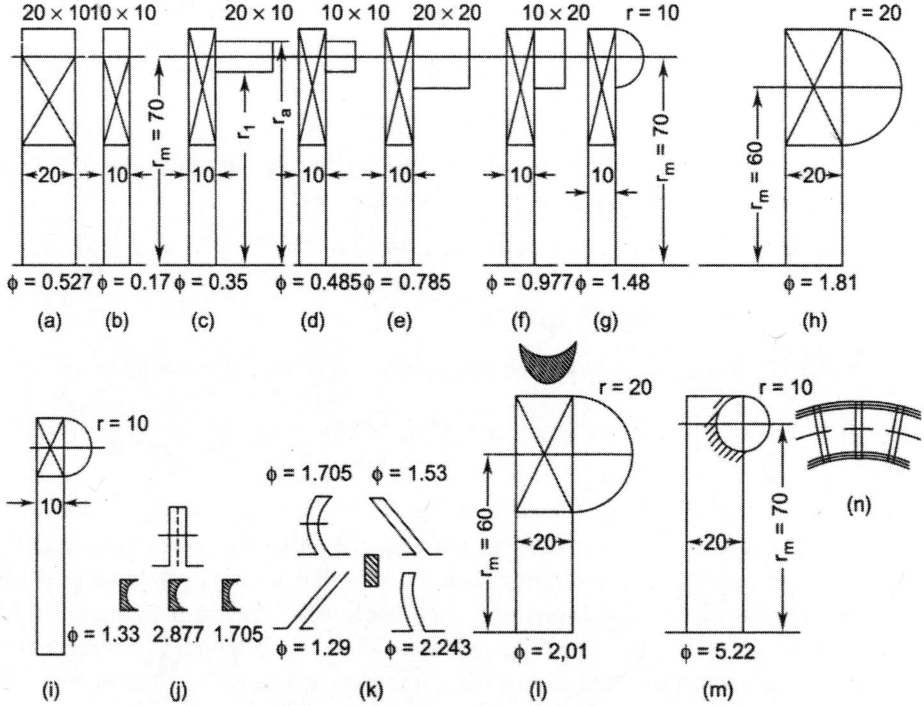

Figure 7.9. Variation of Head co-efficient Ψ with respect to the channel shapes, when (C/u) = 0.5

7.6 CALCULATION OF REGENERATIVE PUMP PARAMETERS

It is found from the experimental results conducted by Prof B.E Nahodhkind /27, 30/, that similarity laws applied for centrifugal pumps well agree for regenerative pumps also. Most of the design parameters for regenerative pumps are taken from model test results of regenerative pumps. The calculations can also be done from the specific speed as done for centrifugal pumps. The following procedure is recommended for the calculation of regenerative pump.

1. The experimental head co-efficient $\psi = \dfrac{H}{(u^2/2g)}$ and efficiency depends upon the dimensions of the impeller and the shape of channel. For the open type vortex

pump (*i.e.*,) regenerative pump, (figure 7.2) ψ = 3.5 to 4.5. Using proper value of ψ, the diameter of the impeller can be determined from the known values of speed and total head.

2. The velocity **C** is taken as **C** = ***u*/2**. For the given flow rate **Q** = **CA**, the area of the side channel is determined.
3. Pump eficfiency is taken from the experimental results.
4. Important parameters are $\frac{b}{c}, \frac{l}{c}, \frac{a}{c}, \frac{L}{l}$. Optimum efficiency depends upon the value of $\frac{C}{u}$ *i.e.*, the velocity at the side channel and blade velocity (figure 8.6*a* and *b*).

Where *c* = channel depth.

b = width of the channel.

A = radial width (at the impeller outlet) of the blade at impeller.

l = the distance between two blades at the middle radius.

L = working length of the channel.

Optimum efficiency and head co-efficient are attained when $\frac{b}{c} \sim 1$ and $\frac{l}{c} \sim 1$.

Optimum value of $\frac{a}{c}$ lies between 0.4 and 0.5.

Open type vortex pumps, regenerative pumps, (Fig. 7.2) give higher efficiency and head than close type vortex pumps, Centripetal pumps,(Fig. 7.1) for the same dimensions

Maximum efficiency is attained when $\frac{C}{u} \sim 0.5$. When $\frac{C}{u}$ decreases, head increases and efficiency drops. In figure 7.10, /1, 18/ non-dimensional characteristics for different side channel dimensions are given. It is seen that, the shut off head increases and HQ curve drops down more steeply, when the radial length of the channel is reduced, for the same value of the channel depth *c*. When the channel depth *c* is increased by two times (*i.e.*, doubled) with respect to the initial value, for the same radial openings of the channel, the flow rate proportionately increases.

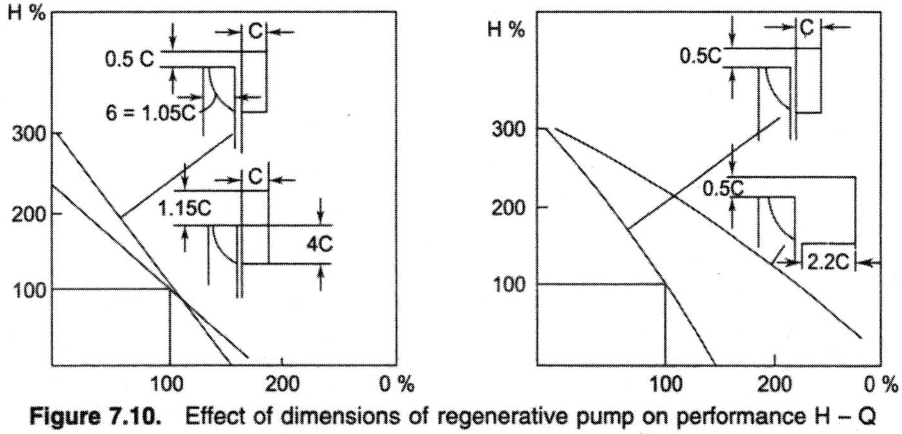

Figure 7.10. Effect of dimensions of regenerative pump on performance H – Q

In figure 7.11(*a*), the effect of number of vanes (Z) on the performance, head and efficiency of the pump is illustrated. The dimensions and the performance are given in figure 7.11(*b*). Total head increases with the increase of number of vanes, considerably at lower flow rate below optimum efficiency point. Overall efficiency mostly remains constant. The form and the shape of the channel considerably influences on head and efficiency. Poor head and efficiency is attained with the channels having straight edges and sharp corners. Rounded corners with straight edges of passage increases the efficiency and the head, higher the radius of curvature of the corners, higher the value of head and efficiency for higher the radius of the rounding, and highest efficiency is attained when the channel shape is semicircle. For easy manufacture, rectangular shape with rounded corners for the channel is selected, since the shut off head and power does not appreciably change in rectangular channels when compared to semi circle channels. However shut off head for both cases *i.e.*, for rectangular and for semicircular shapes is very high.

The cross flow through the clearance between delivery and the suction also influences on the pressure developed. If the 'θ' angle is less, (figure 7.3(*c*)) the channel length 'L' will be more and the total head developed is more, but the leakage from the delivery chamber to the suction chamber through the clearance between the casing and the impeller will also be more. The angle 'θ' must be selected such that the leakage between delivery and suction passage must be minimum. It should be selected in such a way that at least two impeller flow passages must be covered by the casing if not slightly more. Hence it is always better to have more number of impeller passages or in other words more number of blades. This in turn reduces the angle θ. The radial clearance should be maintained minimum in order to reduce the leakage to a minimum. Normal value will be 0.1 mm to 0.15 mm.

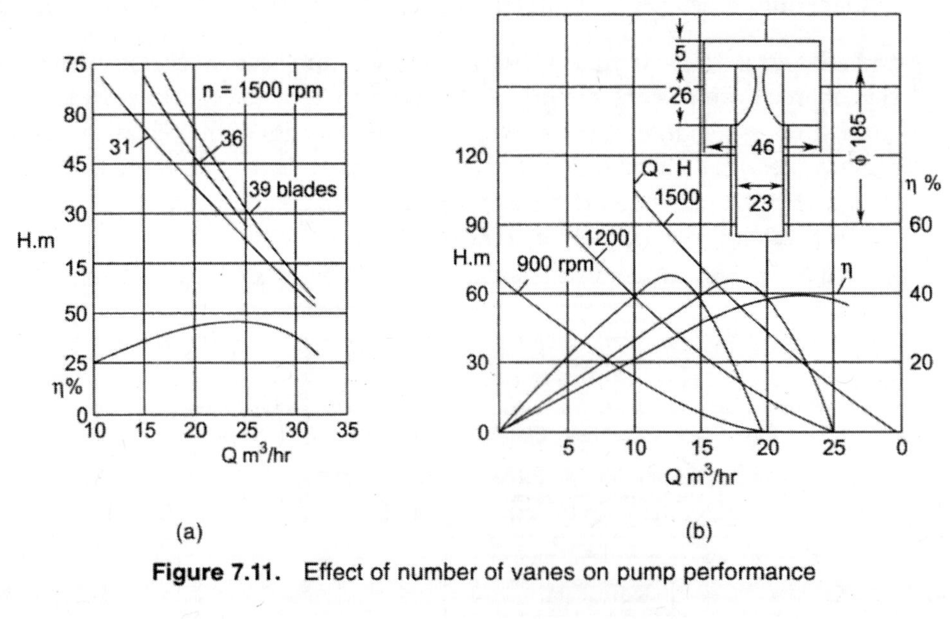

Figure 7.11. Effect of number of vanes on pump performance

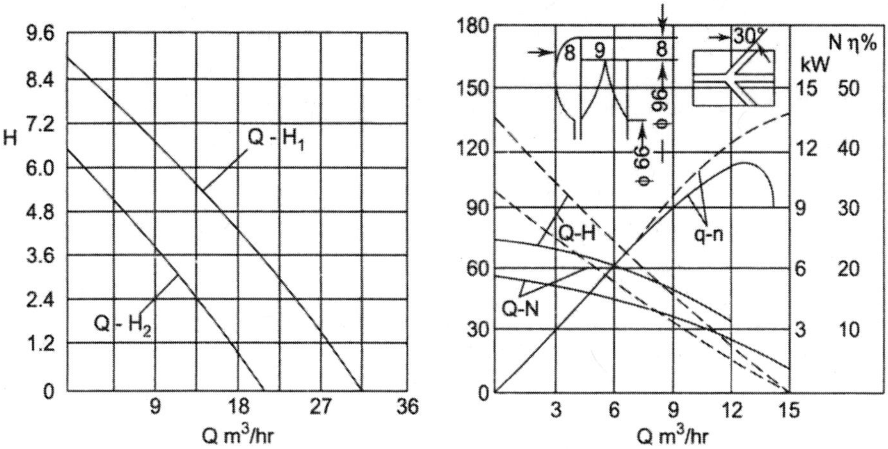

H2-Q for inclined blade impeller
H1-Q for straight bladed impeller

----- line for straight radial blades
——line for inclined blades

Figure 7.12. Comparison between the characteristics of straight bladed and inclined bladed impeller

The form of the blades influence considerably on the performance of the pump. In figure 7.12(*a*)), the pump characteristics $H_2 - Q$ of one regenerative pump having impeller diameter 100 mm with radial blades and running at 3400 rpm is illustrated. In the same graph, the performance of the same pump with inclined radial impeller blades running at 3600 rpm is also given. The pump with inclined blades gives higher head than that of straight blades. The shut off head is also high for inclined impeller blade when compared to straight blades of the pump.

The pump with inclined blades can run at high speeds without any change in the construction. Shut off head also does not increase too high. If the straight blade impeller is to be made for this performance, the diameter of the impeller will be very small and it may not be possible to manufacture. Under such conditions, the inclined bladed impeller is always adopted.

In figure 7.12(*b*), the performance graphs of inclined bladed impeller and straight bladed impeller are illustrated. The performance is better for inclined bladed impeller at very low discharges. The drooping characteristics are less in inclined blades than that in straight blades and shut off head reduces to a greater extent for inclined blade impeller.

7.7 RADIAL THRUST

In regenerative pump, the pressure increases from suction to delivery in radial direction, Due to this, radial thrust is developed, which will be greater in magnitude especially for high-pressure pumps. This radial thrust produces a cantilever force on the shaft and bearings and develops fatigue stress on the shaft. Taking the pressure rise as proportional to radius *i.e.*, straight-line variation, the pressure at the impeller channel passage at any angle measured from the suction port will be (figure 7.13),

$$P\phi = \frac{H}{\phi_0}.\phi \qquad ...(7.21)$$

ϕ_0 – angle subtended for the entire impeller channel between delivery mouth and suction mouth.

θ – angle subtended between delivery mouth and suction mouth, (= 360 – ϕ_0)

b – Width of the impeller.

r – Radius of the impeller.

Force acting on the elementary area under the angle $d\phi$ will be

$$Rd\phi = \frac{H}{\phi_0}.br\phi d\phi \qquad ...(7.22)$$

The resolved components of this force in X and Y directions

$$Rd\phi_x = \frac{H}{\phi_0}.br\phi\cos\phi d\phi$$

and ...(7.23)

$$Rd\phi_y = \frac{H}{\phi_0}.br\phi\sin\phi d\phi$$

Integrating these two forces in order to get, the total force on the entire surface subtending angle ϕ_0

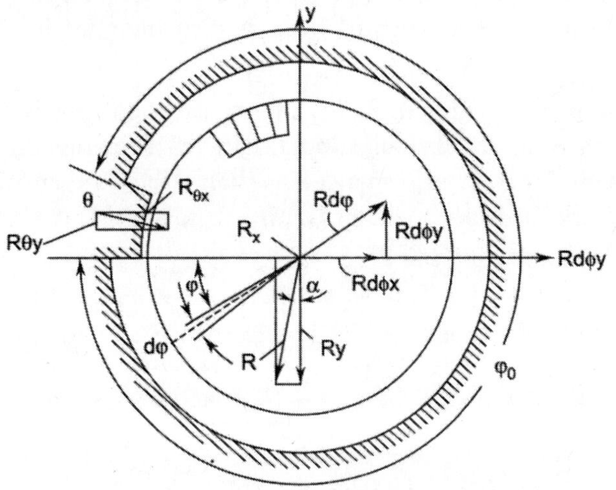

Figure 7.13. Diagram for the calculation of radial forces

$$R_{\phi 0x} = \int_0^{\phi_0} \frac{H}{\phi_0}.br\phi\cos\phi d\phi = \frac{H}{\phi_0}.br\left(\phi_0\sin\phi_0 + \cos\phi_0 - 1\right) \qquad ...(7.24)$$

and

$$R_{\phi 0y} = \int_0^{\phi_0} \frac{H}{\phi_0}.br\phi\sin\phi d\phi = \frac{H}{\phi_0}.br\left(\sin\phi_0 - \cos\phi_0\right)$$

On the surface substituting angle 'θ' (= 360 – ϕ_0) a force acting in X and Y direction

$$R_{\theta x} = \frac{H}{2} . br\theta \cos\frac{\theta}{2} \quad ...(7.25)$$

and

$$R_{\theta y} = \frac{H}{2} . br\theta \sin\frac{\theta}{2}$$

The pressure change in the radial clearance between the delivery mouth and suction mouth will be with respect to time. It can be assumed that the pressure change is in a straight line and the average value will be H/2. The resultant force in X and Y directions will be

$$R_x = \frac{H}{\phi_0} . br\left(\phi_0 \sin\phi_0 + \cos\phi_0 - 1\right) + \frac{H}{2} r\theta.b \cos\frac{\theta}{2} \quad ...(7.26)$$

and

$$R_x = \frac{H}{\phi_0} . br\left(\phi_0 \sin\phi_0 - \cos\phi_0\right) - \frac{H}{2} r\theta.b \sin\frac{\theta}{2}$$

Angles θ and ϕ must be expressed in radians. The magnitude and direction of resultant force will be

$$\tan\alpha = \frac{R_x}{R_y} \text{ and } R = \sqrt{R_x^2 + R_y^2} \quad ...(7.27)$$

For example:

If $\theta = 30°, \phi = 330°$, $R_x = -0.277\,Hrb$, $R_y = -1.046\,Hrb$ and $R = 1.08\,Hrb$ and $\alpha = 14.5°$

Initial pressure will not have any effect on the resultant force. With sufficient accuracy the resultant force can be calculated by the following formula.

$$R = Hrb \quad ...(7.28)$$

The direction will be in Y direction inclined towards low-pressure area. If the blades are located on only one side, axial force will also be prevailing which will be acting eccentrically and with bending force on the shaft.

7.8 PROCEDURE FOR CALCULATION OF REGENERATIVE PUMP

1. The peripheral velocity 'u_2' of the impeller is determined from the known value of H and the Head co-efficient as

$$u_2 = \sqrt{\frac{2gH}{\Psi}}$$

The head co-efficient for closed type units will be Ψ = 4 to 6 and for open type units Ψ = 1.5 to 2.0.

2. The speed is determined from the specific speed 'n_s' which varies from 4 to 50 and from the known value of Q and H as

$$n = \frac{n_s H^{\frac{3}{4}}}{3.65\sqrt{Q}}$$

3. Assuming a vacuum of 4 to 5m, the cavitational property of the pump is checked and a suitable prime mover speed is selected.
4. The tangential velocity 'c_{u2}' with sided channel under specific conditions can be taken as c_{u2} = 0.5 u_2 and Q = 0.59 Q_{max}
5. Area of the side channel determined as $f_3 = f_4 = \dfrac{Q}{C_u\,\eta_0}$. η_0 depends upon the side clearance between impeller an casing and total head.
6. The best shape of side channel is semi circle or nearest to this shape increasing the distance in radial direction with respect to the depth by 2 to 3 times increases the head co-efficient Ψ. If Ψ is higher, the blades will be inclined and sharp β_1 = 65 to 70
7. Axial width of the blade can be selected as equal to the depth of the side channel.
8. Number of vanes can be Z = 24 to 32 for closed channel Z = 12.24 for open channel type units.
9. The tangential to absolute velocity at inner from C_{u2} = 0.5 u_2

7.9 PUMP CHARACTERISTICS

In figure 7.14, the H.Q, N-A characteristic of centripetal pump and in figure 7.15, the characteristics curves for regenerative pump are illustrated. In both these, it is evident that H-Q curves are steep when compared to the curve of centrifugal pump. Hence it is always necessary that the pump must be started only with delivery valve fully opened. In figure 7.16, the performance graph of regenerative pump of 250 mm dia, 10 mm width and (t/c) = 0.4, (figure 7.3(d)), is given. The graph is drawn for different axial and radial clearances and for different suction. It is found that axial clearance, especially at high pressures, considerably influences on the characteristics of the pump. The increase in clearance on one side twice, (from 0.7 mm to 0.15 mm) considerably reduces both total head and discharge. Total head, developed by each stage is almost equal, from a flow rate of 7 m^3/hour, for a suction head of 1.5 m. Total head for the first stage is reduced due to cavitation, which confirms that the first stage must provide sufficient pressure, so that total head increases (dotted lines).

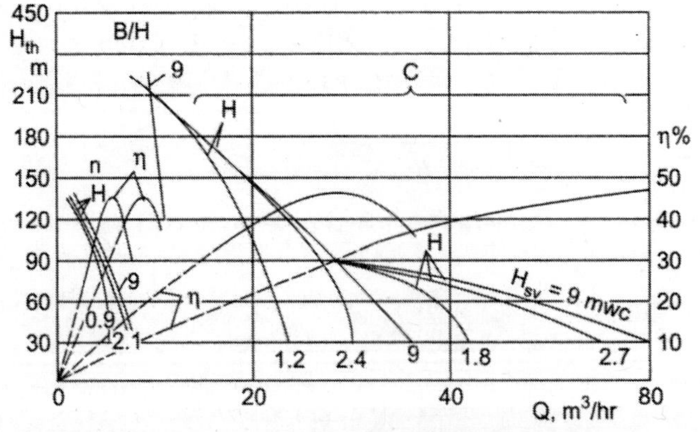

Figure 7.14. Characteristics of centripetal pump (closed impeller)

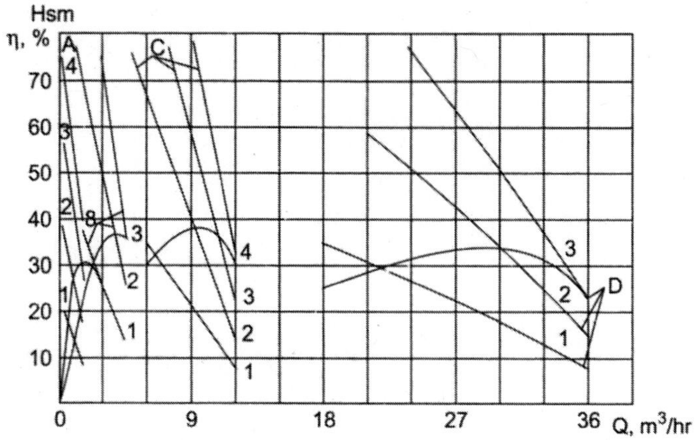

Figure 7.15. Characteristics of regenerative pump (open impeller)

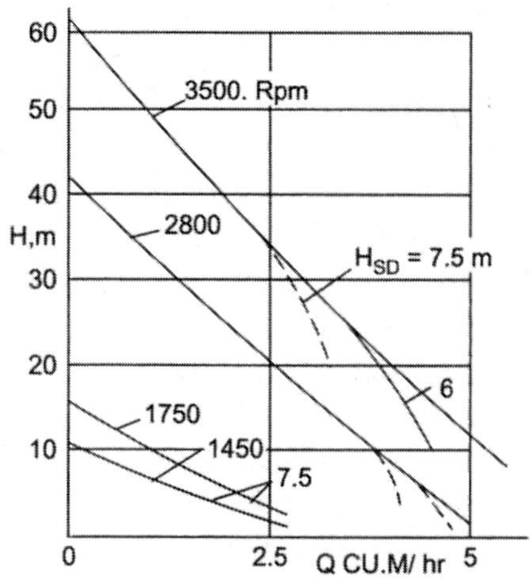

Figure 7.16. Characteristics of regenerative pump at different n and Hs

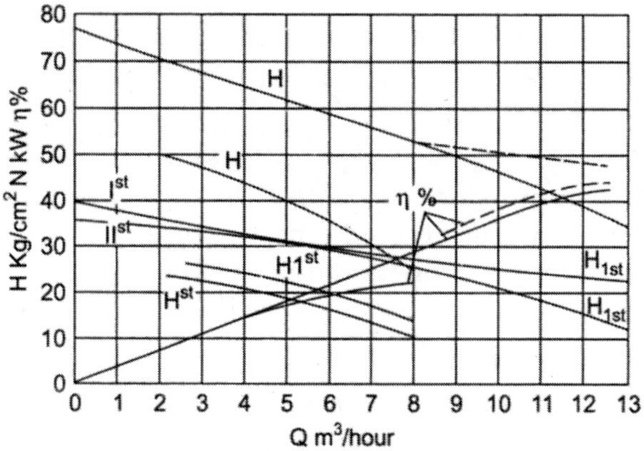

Figure 7.17. Characteristics of a high pressure two stage regenerative pump.

____ Axial clearance 0.075 mm, radial clearance 0.04 mm, suction 1.5 m.

------ Suction 13 m

-.-.- Axial clearance 0.15 mm, Radial clearance 0.115 mm, Suction 1.5 m

CHAPTER **8**

INJECTORS AND EJECTORS (JET PUMPS)

8.1 GENERAL

A jet pump is one, in which, an increase in energy of the pumping fluid, by a working fluid, takes place by a direct contact between both the fluids. The transfer of energy, from the working fluid, to the pumping fluid takes place by direct mixing.

The principle scheme of the jet pump is illustrated in figure 8.1. A jet apparatus, essentially, consists of suction tube 1, suction chamber 4, pressure tube 2, nozzle 3, inlet convergent cone 5, cylindrical mixing tube 6 and a diffuser 7. Working fluid, at a high pressure p_1 atm and at a mass flow rate of 'G_1' kg/s, flows through the pressure pipe 2 and the nozzle 3. At nozzle, the high pressure of the working fluid p_1 is converted into high velocity C_0. The working fluid may be either water or gas or steam or oil. The pumping fluid may be water, oil or sand or dust. The velocity of the fluid coming out of the nozzle, will be approximately 10 to 15 atm, for liquids like water or oil and 15 to 20 ata, for steam and gas. The fluid, to be injected or the pumping fluid, with a very low energy, and at a mass flow rate of G_2 kg/s, enters through the suction tube 1 and suction chamber 4, to the inlet convergent cone 5. Mixing of working fluid and pumping fluid takes place at section 3-3.

The high energy of the working fluid, moving with a velocity, C_0, mixes with the low energy of pumping fluid moving with a very low velocity, C_3. Further movement takes place at the cylindrical section. Velocity C_3 gradually increases and velocity C_0 gradually decreases. At the end of cylindrical tube, at the section X-X, both the fluids, with masses G_1 and G_2 flow together, with a common velocity C_x, due to complete mixing and complete energy transfer between the two fluid.

The length of the cylindrical tube should be properly selected and located away from the nozzle for perfect and complete mixing. At diffuser 7, the kinetic energy of the fluid C_x is converted into the pressure energy. The fluid, at the outlet of the diffuser, at section 4-4, moving with a low velocity, is sufficient enough to flow into the delivery tube, but with high pressure energy.

Jet units work either as injectors or as ejectors. In ejector type jet units, the suction chamber of the unit is connected to another chamber from where the pumping liquid is sucked and delivered to the point where needed. The pumping fluid may be water, gas or oil. The delivery velocity from the diffuser of the jet ejector is low. The jet pump is connected between the suction sump or tank and the suction pipe, in order to increase the suction depth. Taking water from the suction sump and delivering it to the suction pipe of the centrifugal pump by a jet pump is an example.

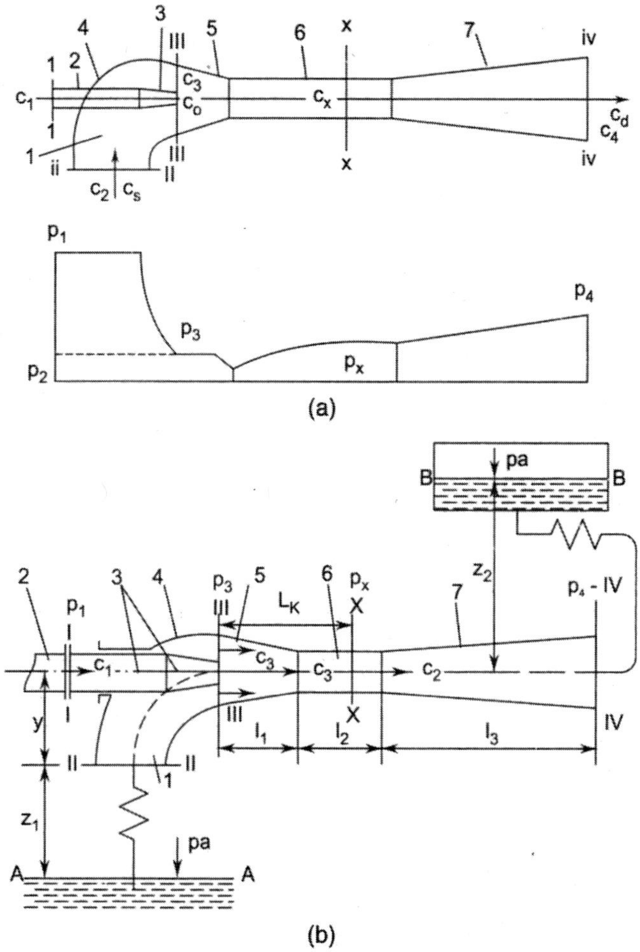

Figure 8.1. Schematic Diagram of a Jet Unit

In injector type jet units, the suction side of the jet unit may be under vacuum or under pressure, but the delivery velocity from the jet pump will be at high velocity. The delivery from the jet injector will be connected to the delivery side of a main line. Connecting the delivery of the jet injector to the delivery pipe of the pump will be an example.

Jet units can be classified as follows :

1. As per the working and pumping fluids such as liquid-liquid unit, gas-gas unit or gas-liquid unit.

2. As per the type of jet apparatus with a two phase flow (steam-water injectors) or single phase flow (water-water ejectors)
3. As per the compression of the mixed gas in jet ventilators and compressors.
4. As per the working process, under vacuum (ejectors) or under pressure (injectors)

Jet pump is one, where in, water-water is used as fluid and operates as ejectors. There is no moving part in jet unit. It is simple in construction, less in weight, possess minimum number of parts and dependable. However the efficiency of the jet pump is low.

Jet units are used, to remove gas in condensers (steam-water ejectors), to remove the dust from the apparatus, like drum (water-water ejectors), supplying water to the boiler (steam injector). Compressibility effect must be considered, while selecting gas or steam units, where as, the density should be considered for water or for any other liquid units.

8.2 WATER JET EJECTORS – JET PUMPS

Water water type ejectors, called jet pumps, work from 0 to 6 m of vacuum and up to a delivery pressure of 10 m of water column. The pressure at the nozzle will be 10 to 16 atm. The water velocity leaving the nozzle will be 40 to 50 m/sec. The pumping liquid, at the ejector, will be 1.5 to 2 m/sec. In the convergent cone, the velocity increases up to 6 to 8 m/sec. The velocity in the mixing cylindrical tube will be 10 to 12 m/sec, and will not be uniform, inspite of providing sufficient length. In diffuser, the velocity reduces up to 3 to 5 m/sec. Flow rate will be 10 to 400 m^3/hr and operating head will be up to 15 m.

8.3 THEORY OF WATER JET PUMP

At the nozzle of water jet pump, the pressure or potential energy is converted into kinetic energy. At the nozzle inlet, at section 1-1, the flow with the potential energy Z, pressure energy p_1 and kinetic energy C_1 is converted into a high kinetic energy C_3 and a low positive pressure p_3. Actual velocity C_0, at section 3-3, can be determined.

Taking the friction coefficient λ_n = 0.05 to 0.1 and since mass flow 'G' remains constant, the energy equation between sections 1 and 3, can be written as

$$\frac{p_1}{\gamma} + \frac{C_1^2}{2g} + Z_1 = \frac{p_3}{\gamma} + \frac{C_0^2}{2g} + \lambda_n \frac{C_0^2}{2g} + Z_0 \quad \text{... (8.1(a))}$$

$$C_0 = \sqrt{\frac{2g\left(\frac{p_1}{\gamma} - \frac{p_3}{\gamma}\right) + C_1^2}{1 + \lambda n}} \quad \text{... (8.1(b))}$$

since $Z_1 = Z_0$

Experiments show, that the velocity C_0 at the nozzle outlet remains constant even up to a distance of 4 to 5 D_n from the nozzle. There after, velocity deformation starts. Axial velocity, at the outer circumference, changes. The velocity distribution is shown in Fig. 8.2. At section A-A of the mixing cylindrical section, mixing takes place with heavy losses. Many authors accept that a turbulent impulse takes place during mixing. Actual pressure and velocity distribution in the chamber indicates that at the initial section 1-1, 2-2, 3-3, the static pressure is unequal and at section 7-7, it is constant.

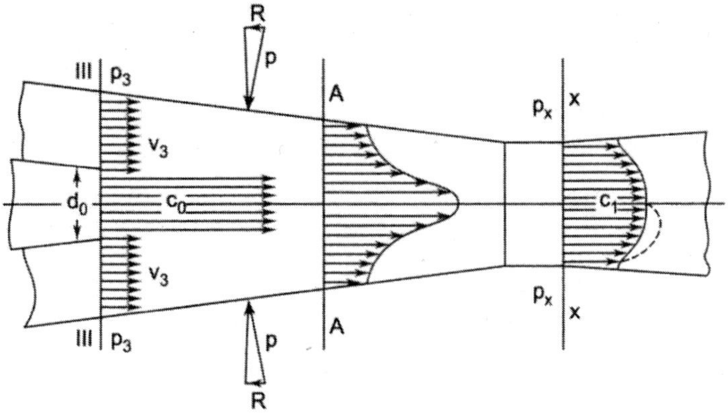

(a) velocity distribution in the mixing chamber of the jet unit

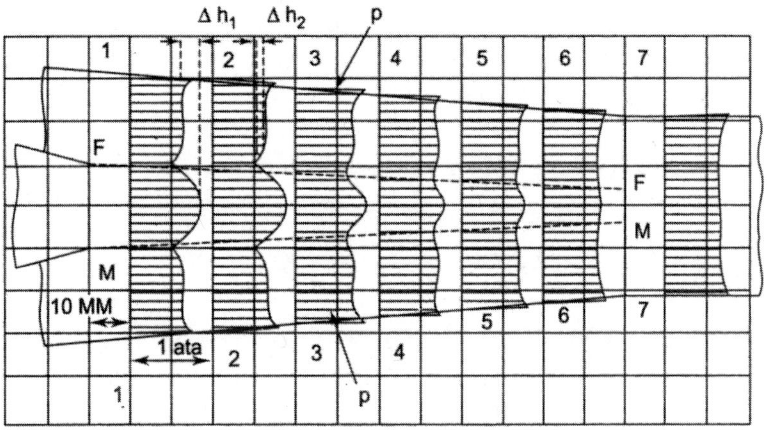

(b) Distribution of static pressure at different sections of mixing chamber

Figure 8.2.

On the surface of the jet stream, at sections F-F and M-M, the pressure p is lower than, the pressure at the surface of the pumping fluid. At a pressure p_1 = 4 to 6 atm, this pressure difference, ΔH_2, is found to be 20 to 50 mwc. The highest pressure difference ΔH_1 is found to be 50 to 150 mwc. At mixing chamber, the pressure is in between jet pressure and pumping fluid pressure. Static pressure remains constant, without much fluctuation.

At the surface of contact of jet stream and pumping fluid stream, intensive fluctuations are observed. Unsteady vortex formation takes place in these areas. These vortices reduce and at section 5-5, 6-6 pressures become equal and constant. The ratio of pumping fluid flow, to the working fluid flow, is called as flow coefficient q.

$$q = \frac{G_2}{G_1}$$

Experiments show, that, highest efficiency is attained, when the pressure in the mixing chamber does not change, for which the length of the mixing chamber L_2 is slightly more than the length L_1 (Fig. 8.1). In that case, the mixing is completed even before the flow reaches the section X-X, *i.e.*, before the entrance to the diffuser. For maximum efficiency condition, the pressure p_x, at section X-X, should be equal to p_3, *i.e.*,. $p_x = p_3$.

For actual calculations, the unevenness in pressure, at section 3-3, can be neglected, *i.e.*, the instantaneous moment of momentum is equal to the average moment of momentum. The same assumption is taken for kinetic energy C_0 and C_3 also *i.e.*, the coefficient β and coriolis coefficient a are equal and is equal to unity. At the end of mixing chamber *i.e.*, at X-X, an unequal distribution of velocity and moment of momentum exist, the coefficient β and α must be considered. In figure 8.2, the velocity distribution is shown. As per V.A Baharoff /14/β = 1.95 and α = 4.42. As per E.A. Edelchik /217/β = 2.45 and α = 7.75. In general, the value β and α widely change, depending upon the geometrical parameters and its working region. It is evident that β and α are higher, if C_x is higher. Such cases arise in high head jet pumps, having a smaller size of mixing chamber length and diameter, *i.e.*, for low value of 'q', due to higher suction lift.

The external force, acting on the pump system, is the force due to friction T between the fluid and the wall of the pump material in the mixing chamber and fluid force R arising due to the unequal area of cross-section III-III and X-X under the pressure $p_3 = p_x$. Instantaneous moment of momentum.

At section III-III $$M_{III} = \frac{G_1}{g} C_0 + \frac{G_2}{g} C_3 \qquad \text{... (8.2)}$$

At the section X-X $$M_X = \left(\frac{G_1 + G_2}{g}\right) \beta C_x \qquad \text{... (8.3)}$$

The change in the moment of momentum, at these sections, per unit time will be equal to the total external force.

$$M_{III} - M_X = T + R \qquad \text{... (8.4)}$$

Assuming forces T and R do not have any influence on other parameters; the equation (8.4) can be rewritten as

$$\frac{G_1}{g} C_0 + \frac{G_2}{g} C_3 = \left(\frac{G_1 + G_2}{g}\right) \beta C_x$$

from which $$q = \frac{C_0 - \beta C_x}{\beta C_x - C_3} \qquad \text{... (8.5)}$$

It is found, from experiments that, at optimum conditions, the value β influences more on the flow co-efficient 'q' than on the forces T and R. Hence, in order to determine q, it is necessary to know the value β or the method to find the same, with the help of other known parameters. The calculations, for other flow conditions, become still more difficult, due to the non- existence of the condition $p_3 = p_x$. Hence 'q' is always determined from experimental results. It is essential to find the frictional losses in mixing cylindrical chamber. In jet pumps, the loss due to vortices, is found to be 70% of total hydraulic loss. The total loss of energy in the mixing chamber will be equal to the difference between the energy at inlet and at outlet. When $p_3 = p_x$ *i.e.*, at optimum conditions

$$\left(\frac{p_{31}}{\gamma}+\frac{C_0^2}{2g}\right)G_1+\left(\frac{p_3}{\gamma}+\frac{C_3^2}{2g}\right)G_2-\left(\frac{p_x}{\gamma}+\frac{\alpha C_x^2}{2g}\right)\times(G_1+G_2)=\Delta E \qquad ...(8.6)$$

This loss will be minimum, if the length of the mixing chamber L_1~ (2.5 to 3.5) d_x and the length of diffuser part L_2 ~ (4 to 8) d_x, where d_x is the diameter of mixing chamber. If L_1 and L_2 are very small full energy does not take place. If L_1 and L_2 are longer, the section X-X shifts from optimum position towards the nozzle and energy transfer is completed earlier in the tube *i.e.*, at the middle of the tube. Further energy conversion from KE to PE takes place at the rest of the length L_i, which again increases the losses. Substituting the experimental values, in the above equation (8.6)

$$\frac{C_0^2}{2g}G_1+\frac{C_3^2}{2g}G_2-\frac{\alpha C_x^2}{2g}(G_1+G_2)=\Delta E \qquad ...(8.7)$$

Eccentricity in nozzle assembly, with respect to the center of mixing chamber, excess length of mixing chamber L_2, improper selection of jet and diffuser inlet diameter lead to higher losses. Conversion of KE into PE takes place in the diffuser part of the jet pump. The equation can be written as

$$\frac{\alpha C_x^2}{2g}(G_1+G_2)=\frac{p_4}{\gamma}(G_1+G_2) \qquad ...(8.8)$$

At the end of the diffuser, section IV-IV, the velocity is C_4. Diffuser losses increase with the increase of the co-efficient α *i.e.*, unevenness in velocity across the cross-section increases. Increasing the length L_2 can attain uniform velocity distribution. Also the coefficient q, will be higher. Hence, the position of diffuser inlet is very important. The diffuser friction coefficient can be selected as $\lambda_{th} \approx 0.1 - 0.3$. Writing down the equation.

$$\left(\frac{p_x}{\gamma}+\frac{\alpha C_x^2}{2g}\right)\times(G_1+G_2)=\left(\frac{p_4}{\gamma}+\frac{C_4^2}{2g}\right)(G_1+G_2)+\lambda_{th}\frac{\alpha C_x^2}{2g}(G_1+G_2) \qquad ...(8.9)$$

Considering the equation (8.9), it is found that, in order to obtain the higher value of p_b, Z_2 or Σh_1, it is necessary to have higher pressure p_4, which is possible only when C_x is higher. But increase in C_x increases β & α *i.e.*, higher losses in the diffuser. That's why, jet pumps are always connected to a high head pumps at suction end or at the throat 4. The hydraulic losses are also less. However, if Z_2 is higher, cavitation may arise at the throat 6 and the pressure p_x will be equal to the vapour pressure of the liquid p_{vp}. To avoid such things, the pressure p_3 will be always selected above p_{vp} approximately by 1 to 1.5 mwc, for which the height Z_2 should be reduced.

Writing down the equation between sections A-A and II – II

$$\frac{p_a}{\gamma}G_2=\left(\frac{p_2}{\gamma}+\frac{C_2^2}{2g}+Z_1+\sum h_1\right)G_2 \qquad ...(8.10)$$

Where C_2 is the average velocity at suction pipe and at section II – II and Σh_1 is the total hydraulic loss.

Energy equation between II – II and III – III is

$$\left(\frac{p_2}{\gamma}+\frac{C_2^2}{2g}\right)G_2=\left(\frac{p_3}{\gamma}+\frac{C_3^2}{2g}+y+\lambda_{th}\frac{C_3^2}{2g}\right)G_2 \qquad \text{... (8.11)}$$

Combining the above two equation (8.10 and 8.11)

$$\frac{p_a}{\gamma}=\frac{p_3}{\gamma}+\frac{C_3^2}{2g}+Z_1+y+\lambda_{th}\frac{C_3^2}{2g}+\sum h_1 \qquad \text{... (8.12)}$$

Taking $y+\lambda_{th}\frac{C_3^2}{2g}$ = Δh dynamic depression at suction

$$H_{sv}=H_{vac.max}+\frac{C_3^2}{2g}-\Delta h \qquad \text{... (8.13)}$$

Taking $\frac{C_3^2}{2g}\approx y$

$$H_{sv}=H_{vac.max}-\lambda_{th}\frac{C_3^2}{2g} \qquad \text{... (8.14)}$$

8.4 ENERGY EQUATION, PARAMETERS AND CHARACTERISTICS OF JET PUMPS

Taking the total energy of the fluid $H_1=\frac{p_1}{\gamma}+\frac{C_1^2}{2g}$ at inlet of the nozzle and $H_2=\frac{p_2}{\gamma}+\frac{C_2^2}{2g}$ for the pumping liquid. $H_4=\frac{p_4}{\gamma}+\frac{C_4^2}{2g}$ for the outlet from the diffuser or at the pressure tube connecting the nozzle, the energy equation will be,

$$H_1G_1=H_4G_2-H_2G_2+yG_2+H_4G_1+\Delta E+ \qquad \text{... (8.15)}$$

$$+\lambda_n\frac{C_0^2}{2g}G_1+\lambda_{th}\frac{C_3^2}{2g}G_2+\lambda_d\frac{\alpha C_x^2}{2g}(G_1+G_2)$$

where $H_1G_1 = N_1$, the input energy of the working fluid.

The Ist three expressions represent the useful energy of the jet pump, working as ejector *i.e.,*

$$N_{ej}=H_{ej}\,.\,G_2 \qquad \text{... (8.16)}$$

But $H_{ej}=H_4-H_2+y$, which is the total head of the ejector or injector. When the pump is working as ejector the 4th expression in equation (8.15) will be the loss in the ejector.

$$H_4\,.\,G_1=\Delta N_{ej} \qquad \text{... (8.17)}$$

If the jet pump works as injector then

$$\Delta N_{inj}=N_{ej}+\Delta N_{ej} \qquad \text{... (8.18)}$$

All other terms, in equation (8.15), are the hydraulic losses at mixing chamber, nozzle, throat, diffuser of the jet pump. Overall efficiency of jet pumps

as ejector,

$$\eta_{ej} = \frac{N_{ej}}{N_1} = \frac{H_{ej}\,G_2}{H_1\,G_1} = q\,\frac{H_{ej}}{H_1} \qquad \text{... (8.19)}$$

as injector

$$\eta_{inj} = \frac{N_{ej}}{N_1} = \frac{N_{ej} + \Delta N_{ej}}{N_1} = q\,\frac{H_{ej}}{H_1} + \frac{H_4}{H_1} \qquad \text{... (8.20)}$$

Dividing equation (8.15) by G_1, and since $q = \frac{G_2}{G_1}$

$$H_1 = qH_{ej} + H_4 + \Delta e + 1_n \frac{C_0^2}{2g} + q\,\lambda_{th}\,\frac{C_3^2}{2g} + (1+q)\,\lambda_d\,\frac{\alpha C_x^2}{2g} \qquad \text{... (8.21)}$$

where

$$\Delta e = \frac{\Delta E}{G_1}$$

It is found from the equation (8.21), that for known value of H_1, and the losses, the change in 'q' will be only due to the change in H_{ej}. Also it is found that, for high pressure, jet pumps will not have high value of 'q'.

For the given value of H_1 and H_{ej} the value of q can be determined from the equation (8.19) and (8.20). Approximate values of q can be determined from the equations (8.5) and (8.21), for which λ_n, λ_{th}, λ_d must be known. From equation 8.9,

$$C_x = \sqrt{\frac{2g\left(H_4 - \frac{p_x}{\gamma}\right)}{\alpha\,(1+\lambda_d)}} \qquad \text{... (8.22)}$$

And from equation (2.14)

$$C_3 = \sqrt{\frac{2g\,(H_{vac.max} - H_{sv})}{\lambda_{th}}} \qquad \text{... (8.23)}$$

Substituting the values of C_3 and C_x in equation (8.5),

$$q = \frac{\frac{1}{\sqrt{1+\lambda_n}}\sqrt{2g\left(\frac{p_1}{\gamma} - \frac{p_3}{\gamma}\right) + C_1^2} - \frac{\beta}{\sqrt{\alpha\,(1+\lambda_d)}}\sqrt{2g\left(H_4 - \frac{p_x}{\gamma}\right)}}{\frac{\beta}{\sqrt{\alpha\,(1-\lambda_d)}}\sqrt{2g\left(H_4 - \frac{p_x}{\gamma}\right)} - \frac{1}{\lambda_{th}}\sqrt{2g\,(H_{vac.max} - H_{sv})}} \qquad \text{... (8.24)}$$

From equation (8.24), it is evident that $q = f\,(\alpha, \beta\,\lambda_c, \lambda_{th}, \lambda_d)$ *i.e.*, depends upon the dimensions and the velocities, exactly Reynolds number. Substituting the values already mentioned earlier, the value of 'C_v', for the given values of H_4, H_{sv}, G_1 and p_1, can be determined from equation (8.24).

It is seen, that for the given value of p, the value 'q' reduces when H_{sv} and H_4 increases individually or both. In figure 8.3,/27,29/ experimental values of $H_s = f\,(q)$ are given for three values of H_1 and H_4. From the graph, it is seen, that when 'q' increases 'H_s' decreases, which is also evident from the equations (8.5), (8.13) and (8.14). This is due to the condition, that the velocity C_3 increases when 'q' increases. Reduction of q, when H_1 is increased, is due to the increased hydraulic losses in mixing chamber and unequal pressure variation at the throat. With the use of equation (8.15) the theoretical $H_{thej} = f(G_{th2})$ can be drawn. For an ejector, equation (8.15) can be approximately written as

$$H_1 G_1 = H_{thej} G_{th2}$$

or

$$H_{thej} = \frac{H_1 G_1}{G_{th2}} \quad \text{... (8.25)}$$

In figure 8.4, graphs are drawn, where $H_1 G_1 = C_1 > C_2 = H_1^1 G_1^1$

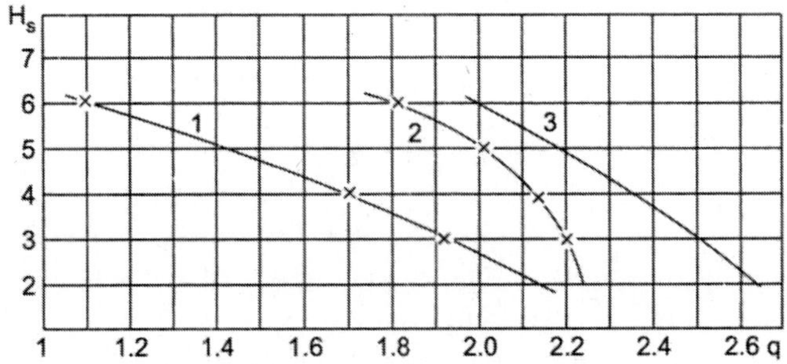

Curve 1 H_1 = 140 mwc and H_4 = 10 mwc

Curve 1 H_1 = 140 mwc and H_4 = 2 mwc

Curve 1 H_1 = 100 mwc and H_4 = 4 mwc

Figure 8.3. $H_s = f(q, H_1, H_4)$

Increase in energy of the working fluid, moves the $H_{thej} - G_{th2}$ curve away from the axis. In figure 8.5, actual $H_3 = f(G_2)$ is also given, which shows that the form of theoretical and actual $H_{thej} - G_{th2}$ curves are same. From the graphs, it is seen that hydraulic losses do not influence on the shape of the curve. Also the inlet pressure increase of the working fluid does not change the shape, but only shifts in parallel away from both axis.

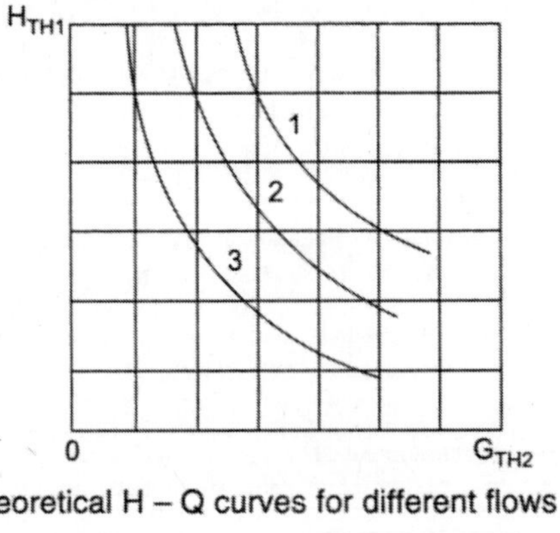

Figure 8.4. Theoretical H – Q curves for different flows through nozzle

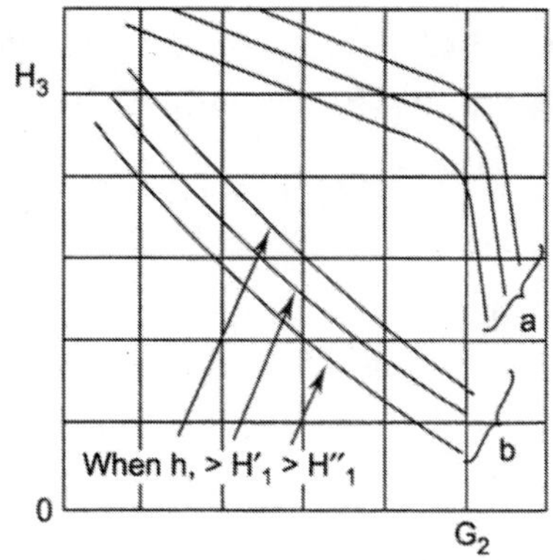

Figure 8.5. Theoretical H – Q curve for different jet pumps

8.5 THEORY

The basic parameters of a jet pump are H_p, the head of working fluid, Q_p, the quantity of working fluid passing through the nozzle, Q, the quantity of the pumping fluid and H, the total head of the pumping fluid. The pumping fluid, after the nozzle, mixes with the working fluid. An energy impulse, *i.e.*, change of energy from the working fluid to the pumping fluid, takes place.

The ratio between the pumping fluid Q and the working fluid Q_p is called, relative flow of the ejector

$$\bar{q} = \frac{Q}{Q_p} \qquad \text{... (8.26)}$$

where q is the relative flow coefficient.

The head developed by the pumping fluid, called as ejector head, H, is

$$H = \frac{p_d - p_s}{\gamma} + (Z_d - Z_s) + \frac{C_d^2 - C_s^2}{2g} + h_f(d-s) \qquad \text{... (8.27)}$$

Where $h_f(d-s)$ is the loss between the suction and the delivery side of the suction pipe, suction chamber, mixing chamber and diffuser. Normally water level in the suction sump will be at the same level of nozzle throat or a little lower than the nozzle throat. The flow velocity in the suction chamber is very small. Hence it is always neglected. $h_f(d-s) \approx 0$, however the velocity, at the section, where the liquid leaves the suction chamber after the nozzle, will be 2 to 2.5 m/sec. So

$$H = \frac{p_d - p_s}{\gamma} + (Z_d - Z_s) + \frac{C_d^2 - C_s^2}{2g} \qquad \text{... (8.28)}$$

where, $\frac{p}{\gamma}$ is the pressure head, Z_d is the potential head, $\frac{C^2}{2g}$ is the velocity head.

Suffix '*d*' refers the outlet, the nozzle mouth and suffix '*s*' refers to the suction tank of the ejector, (figure 8.1). Working fluid, at high head H_p, after leaving the nozzle, mixes with the pumping or ejecting fluid. Energy conversion from KE to PE, takes place by energy transfer, from the working fluid to the pumping fluid,

$$H_p = \frac{p_p - p_d}{\gamma} + (Z_p - Z_d) + \frac{C_p^2 - C_d^2}{2g} + h_f\,(p-d) \qquad \text{... (8.29)}$$

The main loss in the primary line, for the working fluid, is the friction loss in the nozzle cone and in the pipe. Neglecting this loss, since it is very small when compared to the high pressure of operation, the head of working fluid 'H_p' can be written as

$$H_p = \frac{p_p - p_d}{\gamma} + (Z_p - Z_d) + \frac{C_p^2 - C_d^2}{2g} \qquad \text{... (8.30)}$$

where, suffix 'p' refers to the point before nozzle. Since the values

$$\frac{p_p - p_d}{\gamma} >> (Z_p - Z_d) + \frac{C_p^2 - C_d^2}{2g}$$

and

$$\frac{p_d - p_s}{\gamma} >> (Z_d - Z_s) + \frac{C_d^2 - C_s^2}{2g} \qquad \text{... (8.31)}$$

the pumping head can be written as

$$H = H_{st} = \frac{p_d - p_s}{\gamma} \quad \text{and} \quad H_p = H_{pst} = \frac{p_p - p_d}{\gamma} \quad (H_{pst} + H_{st}) = \frac{p_p - p_s}{\gamma}$$

The relative head of ejector is defined as

$$\bar{h}_0 = \frac{H_{st}}{(H_{pst} + H_{st})} = \frac{p_d - p_s}{p_p - p_s} \qquad \text{... (8.32)}$$

The ejector efficiency is the ratio of useful power ($N_0 = \gamma QH$), to the power spent ($N_i = \gamma_p Q_p H_p$).

$$\eta_{ej} = \eta = \frac{N_0}{N_i} = \frac{\gamma}{\gamma_p}\frac{Q}{Q_p}\frac{H}{H_p} = q \cdot \frac{\gamma}{\gamma_p} \cdot \frac{H}{H_p} \qquad \text{.... (8.33)}$$

Efficiency of the ejector depends purely upon the hydraulic losses in the mixing chamber and in the diffuser. Normal values are η = 10 % to 35 %. Referring to figure 8.1 and applying moment of momentum between points 2-2 and 3-3

$$\phi_2 \frac{\gamma Q_p}{g} C_{p2} + \phi_2 \frac{\gamma Q}{g} C_{s2} - \frac{\gamma(Q_p + Q)}{g} C_3 = (p_3 - p_2)\, A_3 \qquad \text{... (8.34)}$$

where C_{p2}, C_{s2}, C_3 are the velocities in the working area before mixing area at section 2-2 after mixing 3-3,

p_3 and p_2 are the pressures at sections 3-3 and 2-2

A_3 is the area of the cylindrical (mixing) section

ϕ_2 is the frictional coefficient for the flow in mixing chamber.

Substituting the values

$$Q = q\,Q_p,\ A_3 = \frac{(1+q)Q_p}{C_3},\ C_{p2} = \phi_1\sqrt{\frac{2g(p_p - p_2)}{\gamma}}$$

$$C_{s2} = \phi_4\sqrt{\frac{2g(p_s - p_2)}{\gamma}},\ C_3 = \frac{1}{\phi_3}\sqrt{\frac{2g(p_d - p_3)}{\gamma}}$$

in equation (8.34) and simplifying,

$$\bar{q} = \frac{\phi_1\phi_2\phi_3\sqrt{p_p - p_2} - k_3\sqrt{p_d - p_2}}{k_3\sqrt{p_d - p_2} - \phi_2\phi_3\phi_4\sqrt{p_s - p_2}}$$

$$\bar{q} = \frac{k_1\sqrt{\Delta p_p + \Delta p_s} - k_3\sqrt{\Delta p_d + \Delta p_s}}{k_3\sqrt{\Delta p_d + p_s} - k_2\sqrt{\Delta p_s}} \quad \text{... (8.35)}$$

where ϕ_1, ϕ_2, ϕ_3 are the velocity coefficients for nozzle, diffuser, and entry to the mixing chamber.

$$k_1 = \phi_1\phi_2\phi_3,\ k_2 = \phi_2\phi_3\phi_4,\ \Delta p_p = p_p - p_s,\ \Delta p_d = p_d - p_s,\ \Delta p_s = p_s - p_2$$

and

$$k_1 = \sqrt{k_p}\left[1 + \frac{\phi_3^2(1-k_p)}{2k_p}\right] \quad \text{and} \quad k_p = \frac{p_d - p_3}{p_d - p_2}$$

The loss is minimum at maximum efficiency point. VTI institute, Moscow/recommends that, at maximum efficiency point

$$\phi_1 = 0.95,\ \phi_2 = 0.975,\ \phi_3 = 0.9,\ \phi_4 = 0.925,\ k_1 = 0.834,\ k_2 = 0.812$$

It is evident from the equation (8.35) that at highest efficiency point, for the given value of p_p, p_d, p_s of the ejector, the value q depends upon the pressures p_2, p_3 or Δp_s and k_3, under the condition $q = q_{max}$. If p_2 = constant, then 'q' as per the equation (8.35), increases with the decrease of the coefficient k_3. Minimum value is attained at the condition $\frac{dk_3}{dk_p} = 0$.

Differentiating the equation with respect k_p and equating it to zero, we get

$$k_p = \frac{\phi_3^2}{2 - \phi_4^2} = 0.67 \quad \text{and} \quad k_3 = 0.98$$

The values p_2 and Δp_s at $q = q_{max}$, condition are determined, while calculating the value q using equation (8.35) for a few values of p_2 in the range. $0 < p_2 < p_s$.

For the given value of p_p and p_s, in the equation (8.35) and for a number of p_d values, optimum value of p_2 and q can be determined, which is used for the calculation of jet pump. Figure 8.6 gives the graphical relation between various parameters as a function of q/27, 30.

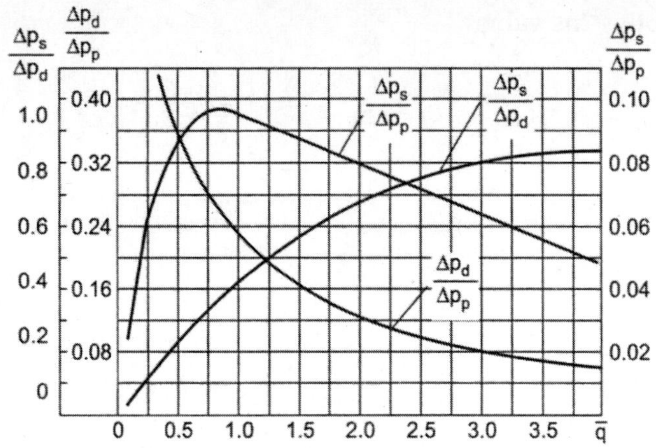

Figure 8.6. Graphs $\frac{\Delta p_d}{\Delta p_p} = f(\bar{q})$; $\frac{\Delta p_s}{\Delta p_d} = f(\bar{q})$; $\frac{\Delta p_s}{\Delta p_p} = f(q)$ for jet pumps at $\eta = \eta_{max}$

8.6 CHARACTERISTICS OF JET PUMPS

The performance of a jet pump is represented by the pumping head H or H_{ST} and efficiency as a function of $Q_{to} = Q + Q_p$, under constant values of Q_p, p_p, and p_s (figure 8.7). The variation of H with respect to Q is $H = f(Q_{to})$ can be obtained from equation (8.34). Substituting the values

$$Q = qQ_p, C_{p2} = \frac{Q_p}{A_n}, C_{s2} = \frac{q\bar{Q}_p}{A_{s2}}, C_3 = \frac{(1+q)Q_p}{A_3}$$

in equation 2.34 and re-arranging

$$\bar{h} = \frac{p_d - p_s}{p_p - p_s} = \phi_1^2 \frac{A_n}{A_3}\left[2\phi_2 + \left(2\phi_2 - \frac{1}{\phi_4} \cdot \frac{A_3}{A_{s2}}\right)\frac{A_n}{A_{s2}} q^2 - \left(2 - \phi_3^2\right)\frac{A_n}{A_3}(1+q)^2\right] \quad ...(8.36)$$

Where A_n is the exit area of the nozzle

$A_{s2} = (A_3 - A_e)$ area at section 2-2 for the pumping liquid

I. A Sokoloff// has given the following expression for the relation between H and 10_{to}.

$$\bar{h} = 2\frac{\phi_1}{\phi_3}\frac{A_n}{A_3}\left[k_1 + \left(k_2 - \frac{\phi_3}{2\phi_1}\right)\frac{A_n}{A_{s2}} q^2 - \left(1 - \frac{\phi_3}{2\phi_1}\right)\frac{A_n}{A_3}(1+q)^2\right] \quad ...(8.37)$$

η and $H - f(Q_{to})$ curves and non dimensional graphs $\bar{h}, \eta = f(q)$ are drawn using equations (8.36) and (8.37), figure 8.7 (*a* and *b*). The non dimensional graphs remain same for similar jet pumps, when working at auto model region, where f remains constant for all values of Reynolds number (R_e). The geometrical similarity of jet pumps is attained if $k_A = \frac{A_3}{A_n}$, is constant, for a series of jet pumps

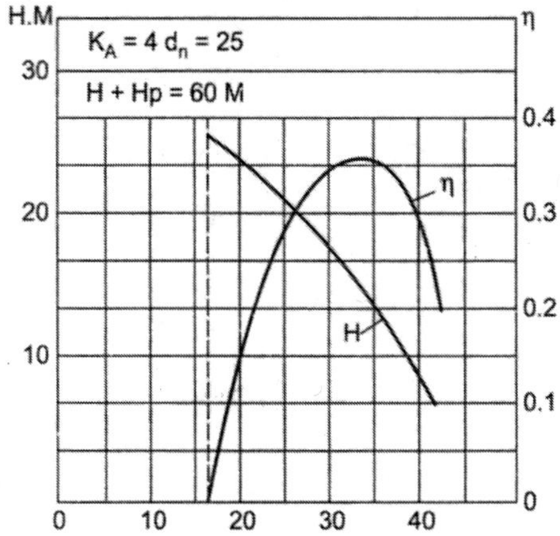

(a) Characteristics of jet pump

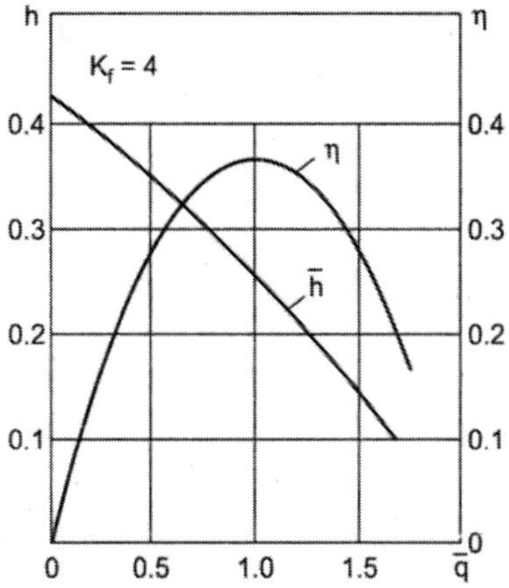

(b) Non dimensional characteristics of a jet pump

Figure 8.7

High head and low discharge ejectors will have a low value of k_A, and k_A increases when the total head decreases and the flow increases. When the discharge (Q) increases the pressure 'p_2' decreases. When $p_2 < p_{vp}$ (vapour pressure), no more pumping can be done due to the presence of cavitation. The critical value of discharge (q_{cr}) can be determined as

$$\bar{q}_{cr} = \left[\frac{A_3}{A_n} - \sqrt{\frac{p_p - p_s}{p_p - p_{vp}}}\right] \cdot \frac{\phi_4}{\phi_1}\sqrt{\frac{p_s - p_{vp}}{p_p - p_s}} \qquad \text{... (8.38)}$$

This critical discharge $\bar{q}_{cr}$ is given for one value of p_s and p_p. When cavitation occurs; further increase in discharge (q) is not possible.

The universal characteristics of jet pump is represented by h and q for various values of (k_A) figure 8.8. All optimum efficiency points of different h-q curves drawn for different k_A values are joined by a smooth curve. This curve represents, the best operating conditions for different jet pumps of different values of k_A. Equation (8.35) represents this curve approximately.

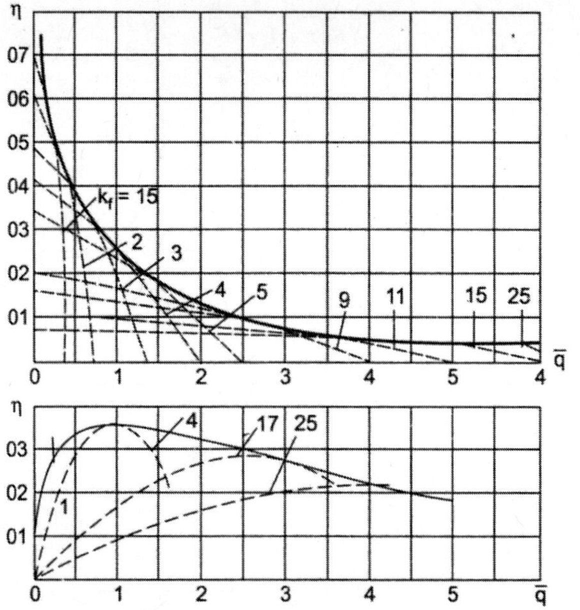

Figure 8.8. Universal characteristics of jet pump

8.7 DESIGN PROCEDURE FOR JET PUMP

The given parameters are the quantity of pumping liquid (Q) working pressure (p_p) before nozzle, the suction pressure before ejector (p_s) and the delivery pressure after jet pump (p_d) or ejector.

1. The relative head $\bar{h}_0$ at optimum condition is determined by

$$\bar{h} = \frac{p_d - p_s}{p_p - p_s} \qquad \text{... (8.39)}$$

2. For the obtained value of $\bar{h}$ the non dimensional coefficient (k_A) is determined for the condition $q = q_{max}$. Equation (8.36).

$$k_A = \frac{A_3}{A_n} = \frac{\phi_1^2 \phi_2}{\bar{h}} \frac{0.88}{h} \qquad \text{... (8.40)}$$

3. For the values of k_f and $\bar{h}_0$ from the universal graph, the value q is determined. Q_0 must be less than q. This must be checked from the equation (8.38)
4. The flow rate of the working fluid (Q_p) is determined as

$$Q_p = \frac{Q}{q_0} \qquad \text{... (8.41)}$$

5. The nozzle diameter (A_n) is determined from the pressure drop ($P_p - P_s$) as

$$A_n = \frac{Q_p}{\phi_1}\sqrt{\frac{\gamma}{2g\left(p_p - p_s\right)}} \quad \text{and} \quad d_n = \sqrt{\frac{4A_n}{\pi}} \qquad \text{... (8.42)}$$

6. The length of free stream from the nozzle exit up to the section where $q = \bar{q}_0$ is $\bar{q}_0 \geq 0.5$ given by

$$l_{n-4} = \frac{0.37 + \bar{q}_0}{4.4a} d_n \quad \text{... (8.43)}$$

Where $a = 0.16$, experimental co-efficient

7. The diameter of the free stream across the length (l_{n-4}) from the nozzle throat exit when $\bar{q}_0 \geq 0.5$ is equal to

$$d_4 = 1.55\left(1+\bar{q}\right) d_n \quad \text{... (8.44)}$$

8. Since $d_4 > d_3$, the mixing chamber will be a convergent cone. The length of the cone will be

$$l_{4-2} = \frac{d_4 - d_2}{2 \tan \dfrac{\gamma}{2}} \quad \text{... (8.45)}$$

Where γ is the angle of convergence $\approx 60°$

9. The distance l_{n-2} from the nozzle exit to mixing chamber inlet $(l_{n-2}) = l_{n-4} + l_{4-2}$ It is recommended to have

$$l_{n-2} = (1.0 \text{ to } 2.0)\, d_3 \quad \text{... (8.46)}$$

The length of the cylindrical section as

$$l_{2-3} = \frac{d_3}{2a_1}\left[1 - \sqrt{\frac{A_n}{A_3}}\left[\frac{0.12q}{\left(k_f - 1\right) + \dfrac{q^2}{\left(h_f - 1\right)} + 0.29}\right]\right] - l_{c-2} \quad \text{... (8.47)}$$

Where $a_1 = 0.05$ the length $l_{2-3} = (6 \text{ to } 10)\ d_3$.

The velocity at the exit from the nozzle is taken as $C_d = 2$ to 5 mps. Based on this the nozzle diameter is

$$d_n = \sqrt{\frac{4\left(1+q\right)Q_p}{\pi C_d}} \quad \text{... (8.48)}$$

Taking the angle of divergence for the diffuser $\varepsilon = 8°$ to 10°, the length of the diffuser passage is

$$l_{3-4} = \frac{d_d - d_3}{2 \tan \dfrac{\varepsilon}{2}} \quad \text{... (8.49)}$$

From equations (8.27) and (8.29), the total head H and H_p can be determined. The efficiency can be determined from equation (8.33). Theoretical performance graphs $H = f(Q_n)$ and $\eta = f(Q_n)$ can be plotted from the established values using equations.

8.8 OPERATION OF CENTRIFUGAL PUMP AND JET PUMP CONNECTED IN SERIES

Jet pumps are connected to centrifugal pumps, when suction height (h_s) is more than the recommended value for the centrifugal pump. Jet pumps are connected at the bottom of the suction pipe of the centrifugal pump. High pressure water required for the working fluid of the jet pump is taken from the delivery side of the centrifugal pump. Certain quantity designated as working quantity (Q_p) always goes round the jet pump. Quantity (Q), the pumping quantity is taken by the jet pump from the suction tank and delivered to the delivery side of the pump.

The centrifugal pump works with the total quality $Q_t = (Q_p + Q)$. In figure 8.9, the schematic diagram of jet – centrifugal pump combination is illustrated. 7 & 1 are suction and delivery pipes of the centrifugal pump. Increase in pressure for the liquid takes place in centrifugal pump 3. A portion of this liquid is sent to nozzle (6) of the jet pump through the pressure pipe 8 from the delivery pipe of the centrifugal pump. This liquid acts as a working fluid and ejects (sucks) the liquid called pumping fluid or ejected fluid or injected fluid from the suction pipe. This pumping fluid passes through the centrifugal pump and delivered to the pump delivery pipe.

If G_p(Kg/sec) is the mass flow, injected (at section IV-IV) from the suction tank of the jet pump, the same mass is delivered to delivery line 1 of the centrifugal pump. (Section II-II).

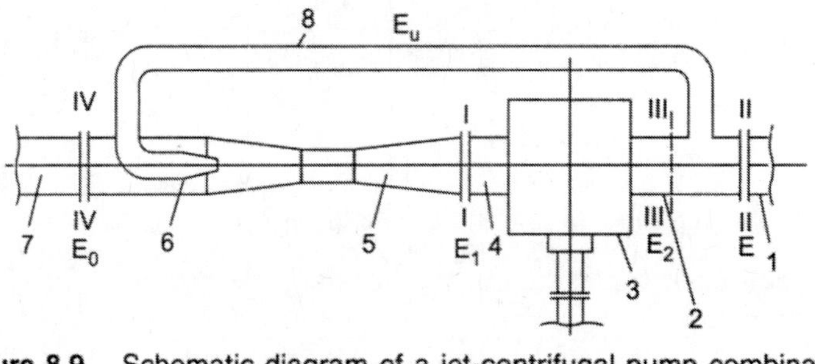

Figure 8.9. Schematic diagram of a jet-centrifugal pump combination

G_w is the mass of the working fluid drawn from the delivery pipe of centrifugal pump and delivered to the nozzle. p and C are the pressure and velocity at this section. Energy of pumping fluid coming out from the section II-II

$$E = G_p\left[\frac{p_2}{\gamma} + \frac{C_2^2}{2g}\right] \qquad \text{... (8.50)}$$

Energy of the working fluid coming out from the section III-III and passing through the pressure pipe (8) and then to nozzle (6) is

$$E_w = G_w\left[\frac{p_2}{\gamma} + \frac{C_2^2}{2g}\right]$$

Total energy will be $E_2 = E_p + E_w = (G_n + G_p)\left[\frac{p_2}{\gamma} + \frac{C_2^2}{2g}\right]$ ... (8.51)

Total energy of the fluid before the pump (section I-I)

$$E_1 = (G_p + G_w)\left[\frac{p_1}{\gamma} + \frac{C_1^2}{2g}\right] \quad \text{... (8.52)}$$

Energy of fluid (section IV-IV) at suction pipe of jet pump

$$E_0 = G_p\left[\frac{p_0}{\gamma} + \frac{C_0^2}{2g}\right] \quad \text{... (8.53)}$$

Net increase in energy of the system *i.e.*, between II-II and IV-IV

$$E_{sy} = E - E_0 \quad \text{... (8.54(a))}$$

And efficiency

$$\eta_{sy} = \frac{E_{sy}}{N} = \frac{E - E_0}{N} \quad \text{... (8.54(b))}$$

Where, N is the power given to the pump shaft. Increase in energy in pump alone *i.e.*, I-I to III-III

$$E_{pump} = E_2 - E_1 \quad \text{... (8.55(a))}$$

And efficiency, $\eta_{pump} = \frac{E_{pump}}{N} = \frac{E_2 - E_1}{N}$ ... (8.55(b))

Efficiency of the injector (or ejector)

$$\eta_w = \frac{E_1 - E_0}{E_N} \quad \text{... (8.56)}$$

Combining all three equations

$$h_{sy} = h_{pump}\frac{E \,.\, E_1 - \eta_w E_w}{E_2 - E_1} \quad \text{... (8.57)}$$

Rearranging the above equation (8.57), by giving the respective values

$$H_{pump} = \left(\frac{p_2}{\gamma} + \frac{C_2^2}{2g}\right) - \left(\frac{p_1}{\gamma} + \frac{C_1^2}{2g}\right)$$ is the pump head.

$$H_2 = \left(\frac{p_2}{\gamma} + \frac{C_2^2}{2g}\right)$$ is the delivery head of the pump

$\frac{H_2}{H_{pump}} = \alpha$ is the head co-efficient of pump

$\frac{G_p}{G_w} = q$ is the flow ratio.

$$\eta_{sy} = \eta_{pump}\left[1 - \frac{(1+\eta_w)\alpha}{1+q}\right] \quad \text{... (8.58)}$$

H is curve, for the jet- centrifugal pump, and is drawn in the same manner as that for a centrifugal pump.

8.9 JET PUMP DESIGN

Given

Flow required, $Q = 50\ m^3/hr$

Pressure at the delivering pipe, $P_4 = 0.8$ atm

Suction height $= H_s = 5$m

Pressure of water before nozzle = 10 atm (gauge)

To determine the quantity Q at the temperature $t_1^0 c = 20° c(= t_2^0 c)$. water density $\gamma_1 = \gamma_2 = 100 \frac{kg}{m^2}$. The pumping liquid is at a pressure $P_a = 760$ mm of H_g.

1. Velocity at the outlet of the nozzle.

From equation (8.1), $C_0 = \sqrt{\dfrac{2g\left(\dfrac{p_1 - p_3}{\gamma}\right) + C_1^2}{1+\lambda_n}}$

Taking λ_n, the frictional co-efficient for the nozzle as 0.07 and $C_1 = 2$ mps, the pressure 'P_3' after the nozzle (at section III-III) is

$$\frac{p_3}{\gamma} = \frac{p_{at}}{\gamma} - hs = \frac{760 \times 13.6}{1000} - 5 = 10.336 - 5 = 5.336 \text{ mwc(abs)} = 05336\text{atm}$$

$$P_1 = 10 + 1 = 11\text{atm(abs)} = 11 \times 10.336 = 113.696 \text{ mwc(abs)}$$

$$C_0 = \sqrt{\frac{2 \times 9.81\,(113.696 - 5.336) + 2^2}{1 + 0.07}} = 44.6 \text{ mps}$$

2. Velocity of the pumping fluid C_3 will be very small, hence selected from the experimental results. C_3 is taken as (0.1 to 0.18) C_o. For this design

$$C_3 = 0.15C_0 = 0.15 \times 44.6 = 6.7 \text{ mps} \approx 6 \text{ mps}$$

3. Pressure 'P_3' at section III-III must be above vapour pressure (P_{vp}), to avoid cavitation. For $t = 20°$ C the vapour pressure is 0.024 kg/cm^2. Adding reserve value of 0.11 to 0.15 atm, the value of $p_3 > p_{vp} + (0.11 \text{ to } 0.15)$ atm

From equations (8.12) and (8.14)

$$p_3 = p_{at} - \gamma h_s - \gamma(1+\zeta_{th})\frac{C_3^2}{2g}$$

$$= 1.0336 - \frac{1000\times 5}{100\times 100} - \frac{1000}{100\times 100}(1+0.2)\frac{6^2}{2\times 9.81}$$

$$= 1.0336 - 0.5 - 0.1825$$

$$= 0.2501\ atm$$

It is found that, $P_3 = 0.3501 > P_{vp} = 0.024 + 0.15 = 0.174$

The suction chamber encircling the nozzle is safe against cavitation. However from the above value it is possible to increases the nozzle velocity, C_3. But increase in Q_2 may increase the temperature t_2 °C which again increases the vapour pressure; Hence the velocity C_3 is kept same.

The frictional coefficient ζ may be taken as equal to the frictional co-efficient for short bend. ζ_{th} is taken as 0.1 to 0.3.

4. Based on the corrected new value of $p_3 = 0.31$ atm, exact velocity C_0 will be

$$C_0 = \sqrt{\frac{2\times 9.81\,(113.696 - 3.2) + 2^2}{1+0.07}} = 45.05\ mps$$

5. The flow ratio q is

$$q = \frac{Q_2}{Q_1} = \frac{\dfrac{1}{\sqrt{1+\zeta_n}}\sqrt{\dfrac{p_1}{\gamma} - \dfrac{p_3}{\gamma} + \dfrac{C_3^2}{2g}} - \dfrac{\beta}{\sqrt{3\beta-2}}\cdot\dfrac{1}{\sqrt{1-\zeta_d}}\sqrt{\dfrac{p_4}{\gamma} - \dfrac{p_x}{\gamma} + \dfrac{C_4^2}{2g}}}{\dfrac{\beta}{\sqrt{3\beta-2}}\cdot\dfrac{1}{\sqrt{1+\zeta_d}}\sqrt{\dfrac{p_4}{\gamma} - \dfrac{p_x}{\gamma} + \dfrac{C_4^2}{2g}} - \dfrac{1}{\sqrt{1+\zeta_{th}}}\sqrt{\dfrac{p_a - p_3}{\gamma} - h_s}}$$

β = the energy co-efficient, and α the velocity coefficient are combined and given as $\alpha \approx 3\beta - 2$. Based on the experimental results, for an unequal distribution of velocity across the flow $\beta = 2$ to 3. Since the vacuum produced at section III-III is small, due to smaller value of P_4 and h_s, β is taken as $\beta = 2$. In case the length of throat is less, β_2 value has to be assumed higher. Coefficient of friction for diffuser is taken as, $\zeta_d = 0.15$ The velocity at the outlet, from the jet pump C_4 is taken as $C_4 = 2$ m/sec. the pressure at the point x in the mixing chamber $p_x = p_3 = 0.31$ atm.

Maximum vacuum ($h_{s\,max}$) is the pressure difference between surface A-A and III-III

$$h_{s\,max} = \frac{p_a - p_{vp}}{\gamma} \approx \frac{p_a - p_3}{\gamma}$$

Since, $p_3 > p_{vp}$, the pump will be safe against cavitation.

Substituting the above values in the equation

$$q = \frac{Q_2}{Q_1} = \frac{\dfrac{1}{\sqrt{1+0.07}}\sqrt{113.696 - 1.3 + \dfrac{2^2}{2\times 9.81}} - \dfrac{2}{\sqrt{3\times 2-2}}\cdot\dfrac{1}{\sqrt{1-0.15}}\sqrt{18 - 3.1 + \dfrac{2^2}{2\times 9.81}}}{\dfrac{2}{\sqrt{3\times 2-2}}\cdot\dfrac{1}{\sqrt{1-0.15}}\sqrt{18 - 3.1 + \dfrac{2^2}{2\times 9.81}} - \dfrac{1}{\sqrt{1+0.2}}\sqrt{\dfrac{1.03 - 0.31}{0.1} - 5}}$$

$$= 2.4$$

6. $Q_1 = \dfrac{Q_2}{\bar{q}} = \dfrac{50}{2.4} = 20.8\,\dfrac{m^3}{\text{sec}}$

7. Jet efficiency (η_{ej}) will be

$$\eta_{ej} = q\frac{H_{ej}}{H_1} = q\frac{\dfrac{p_4}{\gamma}+h_s}{\dfrac{p_1}{\gamma}+\dfrac{C_1^2}{2g}} = 2.4\frac{8+5}{\left(\dfrac{10}{0.1}+10\right)+\dfrac{2^2}{2\times 9.81}} = 0.26 \text{ or } 26\%$$

8. Dimensions of nozzle

A. Throat diameter do $= \sqrt{\dfrac{4Q_1}{\pi C_0}} = \sqrt{\dfrac{4\times 20.8}{\pi\times 45\times 3600}} = 12.8mm \approx 13mm$

B. Length of cylindrical section in nozzle S_0 = (0.25 to 1) d_0
Taking $S_0 = 0.5\,d_0 = 0.5 \times 13 = 6.5mm \approx 7mm$

C. Inlet diameter $d_1 = \sqrt{\dfrac{4Q_1}{\pi C_1}} = \sqrt{\dfrac{4\times 20.8}{\pi\times 45\times 3600}} = 0.0605m \approx 60.5mm \approx 64mm$ as per the standard pipes available.

D. The length of the cone $S_1 = \dfrac{d_1 - d_n}{2\tan\dfrac{\delta}{2}}$

The angle of convergence d = 15 to 20°. Normal value is 17°. Taking δ = 17°

$$S_1 = \frac{64-13}{2\tan 17} = 84.4\text{ mm} = 84\text{ mm}$$

E. Diameter of the suction pipe connecting the suction chamber d_2. Taking c_2= 2m/sec.

$$d_2 = \sqrt{\frac{4Q_2}{\pi C_2}} = \sqrt{\frac{4\times 50}{\pi\times 2\times 3600}} = 0.094\text{ m} \approx 94\text{ mm}$$

However standard pipe size of d_2 = 102 mm (4") is taken

F. The mixing chamber diameter 'd_3' will be

$$\frac{\pi d_3^2}{4} = \frac{Q_1}{C_0} + \frac{Q_2}{C_3}$$

$$d_3 = \sqrt{\frac{4}{\pi}\left[\frac{Q_1}{C_0}+\frac{Q_2}{C_3}\right]} = \sqrt{\frac{4}{\pi}\left[\frac{20.8}{45\times 3600}+\frac{50}{6\times 3600}\right]} \approx 54\text{ mm}$$

$$d_3 = 1.55\left(1+\bar{q}\right)d_n = 1.55\left(1+2.4\right)13 = 46.35 \approx 47\text{ mm}$$

G. The velocity of fluid in the cylindrical mixing chamber C_x will be taking

$\alpha = 3\beta - 2 = (3 \times 2) - 2 = 4$

$$C_x = \sqrt{\frac{2g}{\alpha(1-\zeta_d)q}} \times \sqrt{\left(\frac{p_4}{\gamma} - \frac{p_x}{\gamma} + \frac{C_4^2}{2g}\right)}$$

$$C_x = \sqrt{\frac{2 \times 9081}{4(1-0.15)} \times 18 - 3.1 + 0.204} = 9.4 \frac{\text{m}}{\text{sec}}$$

H. The diameter of the cylindrical section (d_x) will be

$$d_x = \sqrt{\frac{4(Q_1+Q_2)}{\pi C_x}} = \sqrt{\frac{4(Q_1+Q_2)}{\pi C_x}} = \sqrt{\frac{4 \times (20.8+50)}{\pi \times 9.4 \times 3600}} = 0.112 \text{ m} \approx 52 \text{ mm}$$

I. Diameter of the delivery pipe (d_4), taking $C_4 = 2$ m/sec.

$$d_4 = \sqrt{\frac{4(Q_1+Q_2)}{\pi C_4}} = \sqrt{\frac{4 \times (20.8+50)}{\pi \times 9.4 \times 3600}} = 0.0516\text{m} \approx 52\text{mm}$$

$$d_4 = \sqrt{\frac{4(1+\bar{q})Q_1}{\pi C_4}} = \sqrt{\frac{4(1+2.4) \times 20.8}{\pi \times 2 \times 3600}} = 0.112\text{m} = 112\text{mm} \approx 115\text{mm}$$

Considering the standards,

J. The distance between nozzle outlet and mixing cone inlet (II-III)

$$l_{n-3} = \frac{0.37 + q_0}{4.4a} d_n = \frac{0.37 + 2.4}{4.4 \times 0.16} \times 13 = 51.2 \text{ mm} \approx 52 \text{ mm}$$

K. Length of mixing chamber cone (l_1) taking the angle of convergence as 17°

$$l_1 = \frac{d_3 - d_x}{2 \tan 17} = \frac{54 - 52}{2 \tan 17} = 3.27 \text{ mm}$$

Since it is very small the length $l_1 \approx$ (2 to 3) d_x

$l_1 = 2.5 \times 52 = 130$ mm

And the angle a will be

$$\alpha = \tan^{-1} \frac{54 - 52}{2 \times 130} = 0.44° \approx 0.5°$$

So the distance between nozzle & cone entrance = $l_{n-3} + l_1 = 52 + 130 = 182$mm

L. Length of the cylindrical mixing chamber

L_2 = 4 to 7 d_s, sometimes 6 to 10 d_s Taking $l_2 = 5.5 \times 52 = 286$ mm

M. Length of the diffuser (l_3) taking ε = 4 to 8°

Taking $\varepsilon = 5°$

$$l_1 = \frac{d_4 - d_x}{2 \tan \varepsilon} = \frac{115 - 52}{2 \tan 5} = 360 \text{ mm}$$

$$l_x = l_2 = \frac{d_x}{2a_1}\left[1-\sqrt{\frac{1}{k_a}\left[\frac{0.12q}{(k_a-1)+\frac{q^2}{k_a-1}}+0.29\right]}\right]-l_{n-x}$$

a_1 is an experimental coefficient = 0.05 and $k_a = \frac{A_x}{A_n} = \frac{d_x^2}{d_n^2} = \left(\frac{52}{13}\right)^2 = 16$

$$l_2 = \frac{52}{2\times 0.05}\left[1-\sqrt{\frac{1}{16}\left[\frac{0.12\times 2.4}{(16-1)+\frac{2.4^2}{16-1}}+0.29\right]}\right]-182$$

$$l_2 = 478 - 182 = 298 \text{ mm}$$

$$\bar{h} = \frac{\frac{(p_4-p_s)}{\gamma}}{\frac{(p_n-p_4)}{\gamma}} = \frac{0.8\times 10.336+5}{(10+1)10.336-(0.8\times 10.336)} = \frac{13.2688}{105.4273} = 0.1259$$

$$\bar{q} = \frac{Q_p}{Q_w} = \frac{50}{20.8} = 2.4$$

$$q_{cr} = \left[\left[\frac{A_x}{A_n} - \sqrt{\frac{p_p-p_b}{p_p-p_{vp}}}\right]\cdot\frac{\phi_4}{\phi_1}\sqrt{\frac{p_s-p_{vp}}{p_p-p_s}}\right] = \left[\left[\frac{52^2}{13^2} - \sqrt{\frac{10-0.5}{10-0.24}}\right]\cdot\frac{0.925}{0.95}\sqrt{\frac{0.5-0.025}{10-0.5}}\right]$$

$$q_{cr} = 15.8 \text{(safe)}$$

$$\bar{h} = \frac{2\phi_1}{\phi_3}\cdot\frac{A_n}{A_x}\left[k_1 + \left(k_2 - \frac{\phi_3}{2\phi_1}\right)\frac{A_n}{(A_x - A_n)}q^{-2} - \left(1-\frac{\phi_3}{2\phi_1}\right)\frac{A_n}{A_x}(1+q)^2\right]$$

Substituting the values

$$k_1 = 0.834,\ k_2 0.812,\ \phi_1 = 0.95,\ \phi_2 = 0.975,\ \phi_3 = 0.9,\ \phi_4 = 0.925$$

$$\bar{h} = \frac{2\times 0.95}{0.9}\cdot\frac{13^2}{52^2}\left[0.834 + \left(0.812 - \frac{0.9}{2\times 0.95}\right)\frac{13^2}{(52^2-13^2)}q^{-2} - \left(1-\frac{0.9}{2\times 0.95}\right)\frac{13^2}{52^2}(1+q)^2\right]$$

$$\bar{h} = 0.1319\left[0.834 + 0.0186q^{2-} - 0.0337(1+q)^2\right] = 0.1180 + 0.00254_q^{2-} - 0.0045 - 0.0045q^2 - 0.009q$$

$$\bar{h} = -0.00205_q^{2-} - 0.009q + 0.1180$$

$$\bar{h} = 0.094 \text{ for } \bar{q} = 2.4$$

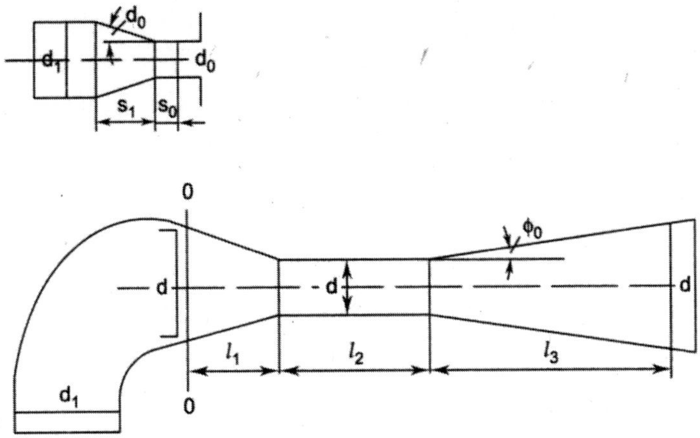

Figure 8.10. Jet pump sketch for the example